Ulrich S. Schubert, Harald Hofmeier,
George R. Newkome

Modern Terpyridine Chemistry

WILEY-VCH

WILEY-VCH Verlag GmbH & Co. KGaA

Authors:

Prof. Dr. Ulrich S. Schubert
Eindhoven University of Technology
Laboratory of Macromolecular Chemistry
and Nanoscience
P.O. Box 513
5600 Eindhoven
The Netherlands

Dr. Harald Hofmeier
Eindhoven University of Technology
Laboratory of Macromolecular Chemistry
and Nanoscience
P.O. Box 513
The Netherlands

George R. Newkome, Ph.D.
The University of Akron
Departments of Chemistry and Polymer Science
Goodyear Polymer Center
170 University Circle
Akron, OH 44325-4717
USA

All books published by Wiley-VCH are
carefully produced. Nevertheless, authors,
editors, and publisher do not warrant the
information contained in these books,
including this book, to be free of errors.
Readers are advised to keep in mind that
statements, data, illustrations, procedural
details or other items may inadvertently be
inaccurate.

Library of Congress Card No.: applied for

British Library Cataloguing-in-Publication Data
A catalogue record for this book is available
from the British Library.

**Bibliographic information published by
Die Deutsche Bibliothek**
Die Deutsche Bibliothek lists this publication in the Deutsche Nationalbibliografie;
detailed bibliographic data is available in
the Internet at <http://dnb.ddb.de>.

© 2006 WILEY-VCH Verlag GmbH & Co.
KGaA, Weinheim

Printed in the Federal Republic of Germany
Printed on acid-free paper

Typesetting   Manuela Treindl, Laaber
Printing   betz-druck GmbH, Darmstadt
Binding   J. Schäffer GmbH, Grünstadt

ISBN-13: 978-3-527-31475-1
ISBN-10: 3-527-31475-X

Preface

Over the past century, synthetic organic chemists have relied more and more on the presence of heteroatoms to realize the ultimate end-goal of their synthetic quest. With the growing importance of metal centers in the supportive infra-structure, it became obvious that the family of *N*-heteroaromatics has become an integral component of this arena. Thus, as our synthetic objectives rely increasingly on pyridine ingredients in these supra(macro)molecular puzzles, we decided that an overview of the syntheses of the parent and substituted 2,2′:6′,2″-terpyridines was in order to lay the foundation for future synthetic endeavors on this interesting group of heterocycles. The use of terpyridine to construct specific, stable, metal complexes will be demonstrated, and their unique properties and assemblies hopefully will inspire others to build on this interesting subunit and to incorporate it as a novel mode of structural connectivity. Although terpyridine was introduced to the synthetic world as early as 1931, it was only after its combination with supramolecular chemistry that its importance was duly realized. Then, its introduction into polymeric assemblies introduced important catalytic properties further emphasizing its importance, expanding new synthetic and nanoscale frontiers. We have attempted to compile the key examples to assist future researchers in this arena. However, although there are many excellent examples in the literature, space constraints have meant that only a limited number of these could be referred to, and we therefore apologize in advance to the authors of research papers that have not been cited.

The authors, as always, would be most grateful to be made aware of any errors which may have crept into the manuscript in spite of the proof-reading that was conducted by many of our acquaintances and colleagues. We also thank spouses, relatives, and friends for their patience and assistance during the completion of this work.

Eindhoven and Akron, January 2006

George R. Newkome
Harald Hofmeier
Ulrich S. Schubert

Modern Terpyridine Chemistry. U. S. Schubert, H. Hofmeier, G. R. Newkome
Copyright © 2006 WILEY-VCH Verlag GmbH & Co. KGaA, Weinheim
ISBN: 3-527-31475-X

Contents

Modern Terpyridine Chemistry. U. S. Schubert, H. Hofmeier, G. R. Newkome
Copyright © 2006 WILEY-VCH Verlag GmbH & Co. KGaA, Weinheim
ISBN: 3-527-31475-X

1
Introduction

Since 1987, when J.-M. Lehn, C. J. Pedersen, and D. J. Cram were honored with the Nobel prize for their results in selective host-guest chemistry [1–3], supramolecular chemistry has become a well-known concept and a major field in today's research community. This concept has been delineated [4] by Lehn: "Supramolecular chemistry may be defined as 'chemistry beyond the molecule', bearing on the organized entities of higher complexity that result from the association of two or more chemical species held together by intermolecular forces." Self-recognition and self-assembly processes represent the basic operational components underpinning supramolecular chemistry, in which interactions are mainly non-covalent in nature (e.g., van der Waals, hydrogen-bonding, ionic, or coordinative interactions); thus, these interactions are weaker and usually reversible when compared to traditional covalent bonds. Nature presents the ultimate benchmarks for the design of artificial supramolecular processes. Inter- and intramolecular non-covalent interactions are of major importance for most biological processes, such as highly selective catalytic reactions and information storage [5]; different non-covalent interactions are present in proteins, giving them their specific structures. DNA represents one of the most famous natural examples, where self-recognition of the complementary base pairs by hydrogen bonding leads to the self-assembly of the double helix. Starting with the development and design of crown ethers, spherands, and cryptands, modern supramolecular chemistry represents the creation of well-defined structures by self-assembly processes [6] (similar to Nature's well-known systems [7]).

One of the most important interactions used in supramolecular chemistry is metal-ligand coordination. In this arena, chelate complexes derived from N-hetero-aromatic ligands, largely based on 2,2'-bipyridine and 2,2':6',2''-terpyridine (Figure 1.1), have become an ever-expanding synthetic and structural frontier.

Bipyridine has been known since 1888 when F. Blau first synthesized a bipyridine-iron complex [8]. One year later, it was again Blau who synthesized and analyzed bipyridine by dry distillation of copper picolinate [9]. Since this parent molecule consists of two identical parts, no directed coupling procedure is required for its construction. Therefore, unsubstituted and symmetrically substituted, in particular 4,4'-functionalized, bipyridines are readily accessible in good yields by simple coupling procedures. Apart from this, bipyridine metal complexes [10] (in

Modern Terpyridine Chemistry. U. S. Schubert, H. Hofmeier, G. R. Newkome
Copyright © 2006 WILEY-VCH Verlag GmbH & Co. KGaA, Weinheim
ISBN: 3-527-31475-X

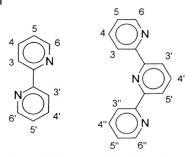

Figure 1.1 Structures of 2,2′-bipyridine and 2,2′:6′,2″-terpyridine.

particular ruthenium complexes) have very interesting photochemical properties making them ideal candidates for solar energy conversion [11].

The chemistry of 2,2′:6′,2″-terpyridines (designated as simply terpyridine or tpy; its other structural isomers are duly noted and will not be considered further here) is much younger than that of 2,2′-bipyridines. In the early 1930s, terpyridine was isolated for the first time by Morgan and Burstall [12, 13], who heated (340 °C) pyridine with anhydrous $FeCl_3$ in an autoclave (50 atm) for 36 h; the parent terpyridine was isolated along with a myriad of other *N*-containing products. It was subsequently discovered that the addition of Fe(II) ions to a solution of terpyridine compounds gave rise to a purple color giving the first indication of metal complex formation. Since this pioneering work was performed, the chemistry of terpyridine remained merely a curiosity for nearly 60 years, at which point its unique properties were incorporated into the construction of supra-molecular assemblies. The number of publications dealing with terpyridine has risen sharply as shown in the histogram (Figure 1.2) – a trend that is predicted to continue, since it is a pivotal structural component in newly engineered constructs based on metallo-polymers and crystal engineering.

The terpyridine molecule contains three nitrogen atoms and can therefore act as a tridentate ligand [14, 15]. It has been extensively studied as an outstanding complexing ligand for a wide range of transition metal ions. The ever-expanding potential applications are the result of advances in the design and synthesis of tailored terpyridine derivatives. The well-known characteristics of terpyridine metal complexes are their special redox and photophysical properties, which greatly depend on the electronic influence of the substituents. Therefore, terpyridine complexes may be used in photochemistry for the design of luminescent devices [16] or as sensitizers for light-to-electricity conversion [17, 18]. Ditopic terpyridinyl units may form polymetallic species that can be used to prepare luminescent or electrochemical sensors [19, 20]. In clinical chemistry and biochemistry, functio-nalized terpyridines have found a wide range of potential applications [21], from colorimetric metal determination [22, 23] to DNA binding agents [24–26] and anti-tumor research [27–29].

Terpyridines have also been utilized for catalytic purposes [30, 31] and in asymmetric catalysis [32]. Another interesting application regarding novel

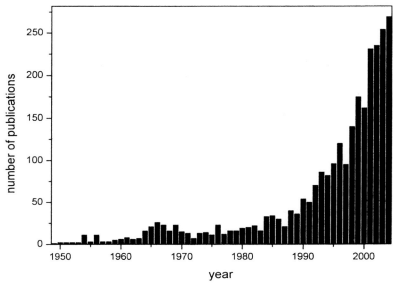

Figure 1.2 Histogram of the number of publications containing the term "terpyridine" using SciFinder (search performed 10.05.2005).

supramolecular architectures is the formation of "mixed complexes", where two differently functionalized terpyridine ligands are coordinated to a single transition metal ion [33–35]. One of the most promising fields for new terpyridine compounds is their unique application in supramolecular chemistry [36]. In this context, the formation of supramolecular terpyridine containing dendrimers [4, 37–41] can be pointed out. Layer-by-layer self-assembly of extended terpyridine complexes on graphite surfaces forms grid-like supramolecular structures [42–45]. Self-assembly of terpyridine compounds on gold [46], CdS [47] or TiO_2 [48], as well as surface functionalization with specially functionalized terpyridine ligands [49], should also be mentioned in this context. Terpyridines, incorporated in macromolecules, enable well-defined supramolecular polymer architectures to be formed, opening up the opportunity of "switching" within physical and chemical properties of materials [34, 35, 50–55].

In view of the notable importance of 2,2':6',2''-terpyridine ligands and their metal complexes in the current research, we herein focus on architectures containing this ligand and the corresponding metal complexes. Therefore, this book is divided into topics featuring different architectures and concepts containing terpyridine metal complexes.

Chapter 2 summarizes the known synthetic strategies leading to different terpyridines. Since 4'-substituted terpyridine currently represents the most valuable family of derivatives, emphasis is directed toward the routes to its synthesis.

Chapter 3 describes the preparation and properties of terpyridine metal complexes. Emphasis will be on *bis*terpyridine-Ru(II) complexes and their optical properties as well as related dyads and triads. Other metal(II) complexes could potentially act as "molecular switches", thus opening up avenues to the construction of nano-devices.

Chapter 4 features various supramolecular aggregates composed of terpyridine-metal subunits, ranging from grids to cyclic structures; moreover, special complexes, where terpyridine complexes are combined with fullerenes or biochemical groups, are described.

Chapter 5 presents polymeric architectures containing terpyridine systems with various architectures, ranging from side-chain-functionalized polymers via main-chain metallopolymers to biopolymers.

Chapter 6 addresses metallodendrimers, micelles, and resins, representing approaches to nanoreactors and immobilized novel catalysts.

Chapter 7 describes catalysis using surface-modified terpyridine metal complexes, opening up potential utilitarian applications, such as assemblies and layers capable of behaving as photoactive materials for use in organic solar cells and LEDs.

References

1 D. J. Cram, *Angew. Chem. Int. Ed. Engl.* **1988**, *28*, 1009–1020.

2 C. J. Pedersen, *Angew. Chem. Int. Ed. Engl.* **1988**, *27*, 1021–1027.

3 J.-M. Lehn, *Angew. Chem. Int. Ed. Engl.* **1988**, *27*, 89–112.

4 G. R. Newkome, R. Güther, C. N. Moorefield, F. Cardullo, L. Echegoyen, E. Pérez-Cordero, H. Luftmann, *Angew. Chem. Int. Ed. Engl.* **1995**, *34*, 2023–2026.

5 J. Darnell, H. Lodish, B. Baltimore, *Molecular Cell Biology*, Scientific American Books, New York, **1990**.

6 J.-M. Lehn, *Supramolecular Chemistry, Concepts and Perspectives*, VCH, Weinheim, **1995**.

7 D. Philp, F. J. Stoddart, *Angew. Chem. Int. Ed. Engl.* **1996**, *35*, 1154–1196.

8 F. Blau, *Ber. D. Chem. Ges.* **1888**, *21*, 1077–1078.

9 F. Blau, *Monatsh. Chem.* **1889**, *10*, 375–380.

10 P. Tomasik, Z. Ratajewicz, *Pyridine Metal Complexes* (Eds.: G. R. Newkome, L. Strekowski), John Wiley & Sons, New York, **1985**.

11 K. Kalyanasundaram, *Coord. Chem. Rev.* **1982**, *46*, 159–244.

12 S. G. Morgan, F. H. Burstall, *J. Chem. Soc.* **1931**, 20–30.

13 S. G. Morgan, F. H. Burstall, *J. Chem. Soc.* **1937**, 1649–1655.

14 W. R. McWhinnie, J. D. Miller, *Adv. Inorg. Chem. Radiochem.* **1969**, *12*, 135–215.

15 E. C. Constable, *Adv. Inorg. Chem. Radiochem.* **1986**, *30*, 69–121.

16 A. Harriman, R. Ziessel, *Coord. Chem. Rev.* **1998**, *171*, 331.

17 B. O'Regan, M. Grätzel, *Nature* **1991**, *353*, 737–740.

18 O. Kohle, S. Ruike, M. Grätzel, *Inorg. Chem.* **1996**, *35*, 4779–4787.

19 M. Schmittel, H. Ammon, *Chem. Commun.* **1995**, 687–688.

20 M. T. Indelli, C. A. Bignozzi, F. Scandola, J.-P. Collin, *Inorg. Chem.* **1998**, *37*, 6084–6089.

21 B. N. Trawick, A. T. Daniher, J. K. Bashkin, *Chem. Rev.* **1998**, *98*, 939–960.

22 B. Zak, E. S. Baginsky, E. Epstein, L. M. Weiner, *Clin. Chim. Acta* **1970**, *29*, 77–82.

23 B. Zak, E. S. Baginsky, E. Epstein, L. M. Weiner, *Clin. Toxicol.* **1971**, *4*, 621–629.

24 H. Q. Liu, T. C. Cheung, S. M. Peng, C. M. Che, *Chem. Commun.* **1995**, 1787–1788.

25 P. M. V. Vliet, M. S. Toekimin, J. G. Haasnoot, J. Reedijk, O. Novakova, O. Vrana, V. Brabec, *Inorg. Chim. Act.* **1995**, *231*, 57–64.

26 P. J. Carter, C. C. Cheng, H. H. J. Thorp, *J. Am. Chem. Soc.* **1998**, *120*, 632–642.

27 W. C. Xu, Q. Zhou, C. L. Ashendel, C. T. Chang, C. J. Chang, *Bioorg. Med. Chem. Lett.* **1999**, *9*, 2279–2282.

28 D. S. H. L. Kim, C. L. Ashendel, Q. Zhou, C. T. Chang, E. S. Lee, C. J. Chang, *Bioorg. Med. Chem. Lett.* **1998**, *8*, 2695–2698.

29 Y. Zhang, C. B. Murphy, W. E. Jones, *Macromolecules* **2002**, *35*, 630–636.

30 K. Umeda, A. Nakamura, F. Toda, *Bull. Chem. Soc. Jpn.* **1993**, *66*, 2260–2267.

31 A. Vavasori, L. Toniolo, *J. Mol. Cat. A: Chemical* **2000**, *151*, 37–45.

32 G. Chelucci, A. Saba, F. Soccolini, D. Vignola, *J. Mol. Cat. A: Chemical.* **2002**, *178*, 27–33.

33 E. C. Constable, C. E. Housecroft, T. Kulke, C. Lazzarini, E. R. Schofield, Y. Zimmermann, *J. Chem. Soc., Dalton Trans.* **2001**, 2864–2871.

34 U. S. Schubert, P. R. Andres, H. Hofmeier, *Polym. Mater.: Sci. Eng.* **2001**, *85*, 510–511.

35 M. Heller, U. S. Schubert, *Macromol. Rapid Commun.* **2002**, *23*, 411–415.

36 J. P. Sauvage, J. P. Collin, J. C. Chambron, C. S. Guillerez, Coudret, V. Balzani, F. Barigelletti, L. D. Cola, L. Flamigni, *Chem. Rev.* **1994**, *94*, 993–1019.

37 U. S. Schubert, C. H. Weidl, C. N. Moorefield, G. R. Baker, G. R. Newkome, *Polym. Prepr.* **1999**, *40*, 940–941.

38 H.-F. Chow, I. Y.-K. Chan, P.-S. Fung, T. K.-K. Mong, M. F. Nongrum, *Tetrahedron* **2001**, *57*, 1565–1572.

39 D. J. Diaz, S. Bernhard, G. D. Storrier, H. D. Abruña, *J. Phys. Chem. B* **2001**, *105*, 8746–8757.

40 G. R. Newkome, F. Cardullo, E. C. Constable, C. N. Moorefield, A. M. W. C. Thompson, *Chem. Commun.* **1993**, 925–927.

41 G. R. Newkome, E. He, *J. Mater. Chem.* **1997**, *7*, 1237–1244.

42 G. S. Hanan, D. Volkmer, U. S. Schubert, J.-M. Lehn, G. Baum, D. Fenske, *Angew. Chem. Int. Ed. Engl.* **1997**, *36*, 1842–1844.

43 A. Semenov, J. P. Spatz, M. Möller, J.-M. Lehn, B. Sell, D. Schubert, C. H. Weidl, U. S. Schubert, *Angew. Chem. Int. Ed.* **1999**, *38*, 1021–1027.

44 U. Ziener, J.-M. Lehn, A. Mourran, M. Möller, *Chem. Eur. J.* **2002**, *8*, 951–957.

45 D. G. Kurth, M. Schütte, J. Wen, *Colloids and Surfaces A: Physicochem. Eng.* **2002**, *198–200*, 633–643.

46 L. S. Pinheiro, M. L. A. Temperini, *Surface Sci.* **2000**, *464*, 176–182.

47 G. Billancia, D. Wouters, A. A. Precup, U. S. Schubert, *Polym. Mater.: Sci. Eng.* **2001**, *85*, 508–509.

48 C. R. Rice, M. D. Ward, M. K. Nazeeruddin, M. Grätzel, *New J. Chem.* **2000**, *24*, 651–652.

49 M. Maskus, H. D. Abruña, *Langmuir* **1996**, *12*, 4455–4462.

50 U. S. Schubert, C. Eschbaumer, *Angew. Chem. Int. Ed.* **2002**, *41*, 2892–2926.

51 M. Heller, U. S. Schubert, *Macromol. Rapid Commun.* **2001**, *22*, 1358–1363.

52 J.-F. Gohy, B. G. G. Lohmeijer, U. S. Schubert, *Macromolecules* **2002**, *35*, 7427–7435.

53 J.-F. Gohy, B. G. G. Lohmeijer, U. S. Schubert, *Macromol. Rapid Commun.* **2002**, *23*, 555–560.

54 U. S. Schubert, H. Hofmeier, *Macromol. Rapid Commun.* **2002**, *23*, 561–566.

55 J.-F. Gohy, B. G. G. Lohmeijer, S. K. Varshney, U. S. Schubert, *Macromolecules* **2002**, *35*, 4560–4563.

2
Syntheses of Functionalized 2,2′:6′,2″-Terpyridines

2.1
Introduction

In view of the wide range of research and potential utilitarian applications of 2,2′:6′,2″-terpyridines, an easily accessible "pool" of different functionalized building blocks is mandated. Therefore, highly efficient routes to these ligands are as essential as their well-defined derivatization at every ring position. Functional groups may be introduced directly during their construction or by various substitution interconversions. While publications concerning the chemistry of terpyridine complexes continued to increase, comparatively few preparations of functionalized 2,2′:6′,2″-terpyridine ligand derivatives have been reported as yet. In 1997, Cargill-Thompson [1] reviewed the historical syntheses of the simple terpyridine ligands, and in 2003 Fallahpour [2] reviewed the 4′-substituted terpyridines. In this chapter, both innovative new synthetic strategies and an up-to-date overview of the classical approaches leading to new 2,2′:6′,2″-terpyridine derivatives will be presented. The new developments in the preparation of chiral terpyridines [3–6] and ditopic terpyridine containing ligands [7–14] can be found elsewhere.

2.2
Basic Synthetic Strategies

The two basic synthetic approaches to terpyridines are by either central ring-assembly or coupling methodologies. Ring assembly is still the most prevalent strategy, but because of their multiplicity and efficiency, modern Pd-catalyzed, cross-coupling procedures have recently become seriously competitive and may surpass the traditional ring-closure processes.

2.2.1
Ring Assembly

Over the last couple of decades, various new terpyridine ring-assembly strategies have been developed; Scheme 2.1 displays these frequently used routes. The most

Modern Terpyridine Chemistry. U. S. Schubert, H. Hofmeier, G. R. Newkome
Copyright © 2006 WILEY-VCH Verlag GmbH & Co. KGaA, Weinheim
ISBN: 3-527-31475-X

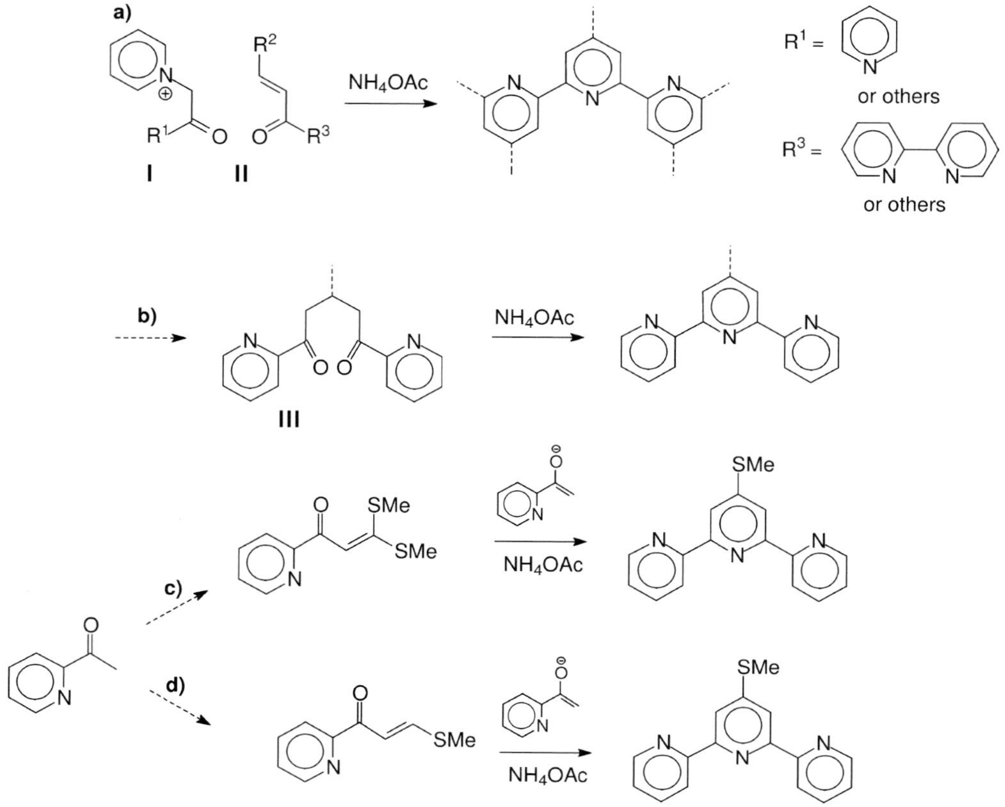

Scheme 2.1 Ring-assembly methods to terpyridines:
(a) and (b) Kröhnke reaction, (c) Potts methodology, (d) Jameson methodology.

common ring assembly of terpyridines is still the well-known Kröhnke condensation (Route a), which initially involves synthesis of *N*-heteropyridinium salts,
e.g., **I**, then subsequent ammonia condensation with an enone **II** [15, 16]. Other
important methods are the initial construction of 1,5-diketones **III** and subsequent
ring closure with an appropriate *N*-source (Route b) [17-19], α-oxoketene dithio-
acetal methodology (Route c) [16], and the Jameson method by condensation of
an *N,N*-dimethylaminoenone with 2-acetylpyridinenolate (Route d) [20].

The major disadvantage of these methods is that the final condensation step
usually yields tarry crude by-products that require special efforts to isolate and
purify the desired terpyridine.

High yield conversions with good product purities were obtained by a four-step
procedure starting from the commercially available 2,6-diacetylpyridine (**IV**), which
was subsequently converted to the 2,6-*bis*(*N*-cyclohexylacetimidoyl)pyridine (**V**),
derived from a multistep procedure, by reaction with cyclohexylamine (Scheme 2.2)
[21]. Cyclization of **V** with *Si*-protected 3-bromopropylamines **VI** afforded the

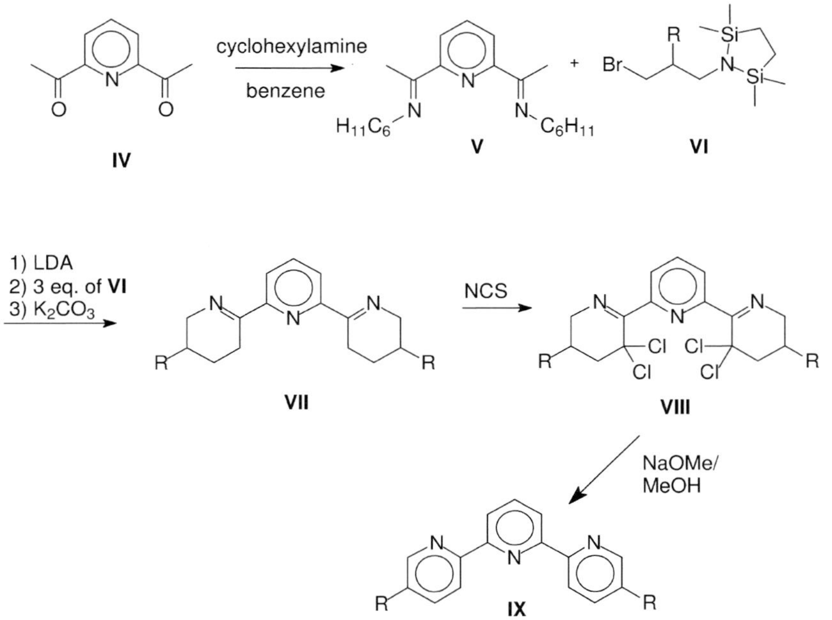

Scheme 2.2 Terpyridine synthesis from 2,6-diacetylpyridine.

tetrahydropyridines **VII**, which, after chlorination, gave the tetrachloro adducts **VIII**, then onto the desired terpyridines **IX** with a respectable (73–93%) overall yield from **V**.

An effective and simple two-step Kröhnke-type [15] synthesis of polysubstituted symmetric terpyridines from 2,6-diacetylpyridine (**IV**) has been described by Sasaki et al. [22] (Scheme 2.3), in which *bis*pyridinium iodide **X**, obtained (85%) from **IV**, was subsequently reacted with various α,β-unsaturated aldehydes **XI** at 80 °C for 4 h in formamide in the presence of ammonium acetate to give in variable yields the different symmetric terpyridines **XII**.

A novel 4-functionalized 2,6-diacetylpyridine **XV**, the key intermediate for the Kröhnke methodology [15], was prepared from 4′-hydroxy-2,6-pyridinedicarboxylic acid **XIII** (Scheme 2.4) [23], by initial esterification to give diethyl 4-chloropyridine-2,6-dicarboxylate, which yielded the corresponding diacid; subsequent conversion to the 2,6-*bis*(chlorocarbonyl)-4-ethoxypyridine (**XIV**) was accomplished. The

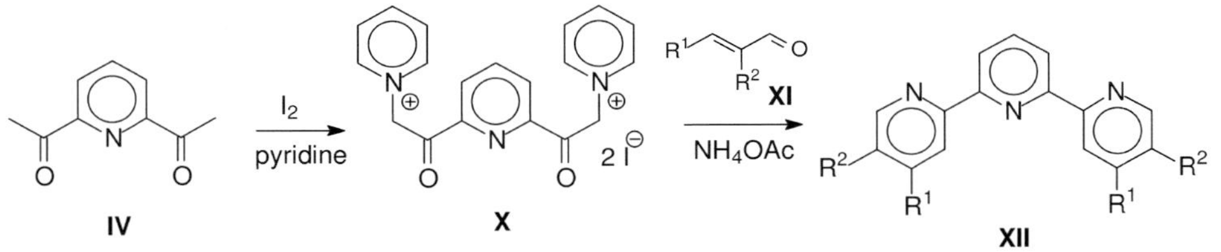

Scheme 2.3 Sasaki-type Kröhnke reaction on 2,6-diacetylpyridine.

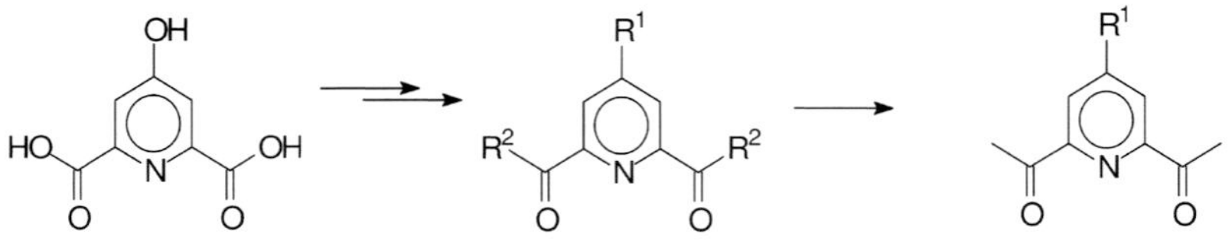

Scheme 2.4 Synthetic approach to functionalized 2,6-diacetylpyridines, as intermediates for the preparation of terpyridines.

reaction with 2,2-dimethyl-1,3-dioxan-4,6-dione followed by hydrolysis with aqueous acetic acid resulted in the formation (36%) of 4-ethoxy-2,6-diacetylpyridine (**XV**). Subsequent Kröhnke-type procedures afforded the desired 4'-substituted terpyridine derivatives.

Recently, the Kröhnke method has also been modified to yield terpyridines under solventless conditions; thus, grinding the starting materials with solid NaOH leads to the quantitative formation of the diketone within 20 min [24–26]. Besides the fast and facile reaction procedures, the environmental friendliness (no solvents are used) is a main advantage of this useful modification.

2.2.2
Cross-Coupling Procedures

In the last few years, appropriate methodologies for the construction of functionalized terpyridines were based on directed cross-coupling procedures. Traditional examples, such as the cross-coupling of organosulfur compounds [27] or lithiopyridines with CuCl$_2$ [28], have the disadvantage that they generally result in overall poor conversions. Modern Pd(0)-catalyzed coupling reactions combine the desired efficiency and simplicity with controllable substitution possibilities. Suzuki [29], Negishi [30], and Stille couplings [31] are all based on a Pd(0)/Pd(II) catalytic cycle. Particularly, the Stille cross-coupling has become a popular route to terpyridines, because of its (a) universal building block principle, (b) multi-gram product accessibility, and (c) well-directed functionalization at almost every desired position (Scheme 2.5) [32–35]. 2,2':6',2''-Terpyridines, functionalized at the central

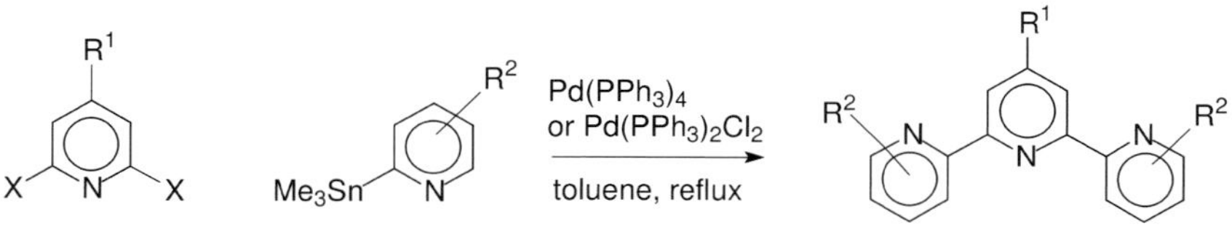

X = Cl, Br, I

Scheme 2.5 Stille-coupling 2-trimethylstannylpyridines and 2,6-dihalopyridines.

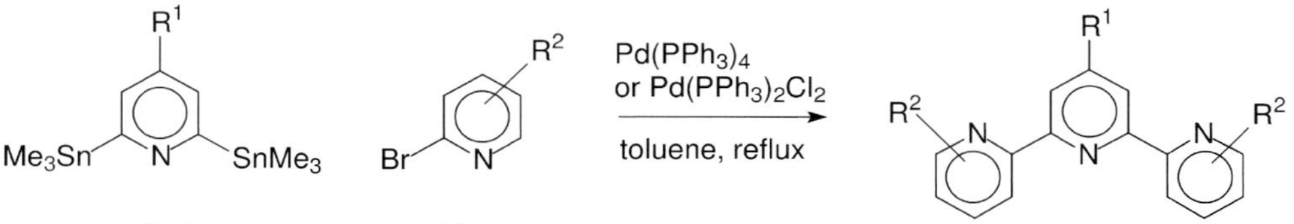

Scheme 2.6 Stille-coupling of *bis*(trimethylstannyl)pyridines and 2-bromopyridines.

and/or terminal pyridine rings, can be obtained utilizing appropriate 2,6-dihalo-pyridines **XVI** as central building blocks, which can be reacted with 2-trialkyl-stannylpyridines **XVII** and Pd(0) catalysts in toluene for at least 24 h.

Terpyridine synthesis via the Stille procedure can be conducted by utilizing 2,6-*bis*(trimethylstannyl)pyridines **XVIII** as a central ring, and coupling them with the corresponding 2-bromopyridines **XIX** (Scheme 2.6) [36, 37].

Other Pd-catalyzed cross-coupling procedures have not yet been used for the synthesis of terpyridines themselves, but seem to be appropriate methods; for instance, Negishi cross-coupling was used for the synthesis of terpyridine-related compounds [38] and related 2,2′-bipyridines [39] in excellent yields.

2.3
Synthesis of 2,2′:6′,2″-Terpyridine Derivatives

Terpyridines may be functionalized at both the central and the terminal rings; therefore, the desired groups must be incorporated into the initial substituted starting compounds via ring-assembly or coupling procedures. In this overview, the terpyridine derivatives are organized by their ring-substitution positions.

2.3.1
4′-Substituted-2,2′:6′,2″-Terpyridinoxy Derivatives

4′-Terpyridinoxy derivatives represent a dominant substitution pattern because of their convenient accessibility via (a) nucleophilic aromatic substitution of 4′-halo-terpyridines by any primary alcohols and analogs or (b) S_N2-type nucleophilic substitution of the alcoholates of 4′-hydroxyterpyridines (the "enol" tautomer of the 4-terpyridone). An overview of the routes is presented in Scheme 2.7. A large variety of functional terpyridinoxy derivatives have been originated from these methods (Table 2.1).

Sampath et al. [40] and Schubert et al. [41, 42] reported a number of linear 4′-terpyridinyl-ethers with terminal hydroxy- (**1a–e**), carboxy- (**1f–g**) (see also [43]), *tert*-butoxy- (**1h**), thio- (**1i**) and amino-groups (**1j–k**). These ethers were prepared in high (~60–90%) yields from 4′-chloroterpyridine with an alcohol and a suspension of base (KOH or NaH) in polar non-protic solvent (DMSO or DMF). The same route was utilized to synthesize 4′-(3-phenylpropoxy)terpyridine (**1l**) [44].

Table 2.1 4'-Terpyridinoxy derivatives.

No./Lit.	R	No./Lit.	R
1a [40]	$(CH_2)_3OH$	**1j** [40]	$(CH_2)_3NH_2$
1b [40, 41]	$(CH_2)_4OH$	**1k** [41, 42]	$(CH_2)_5NH_2$
1c [40, 41]	$(CH_2)_6OH$	**1l** [44]	$(CH_2)_3Ph$
1d [40]	$(CH_2)_8OH$	**1m** [45]	
1e [40]	$(CH_2)_{10}OH$	**1n** [46, 47]	$-C{\equiv}CH$
1f [40, 43]	$(CH_2)_3CO_2H$	**1o** [47]	n = 1, 3
1g [41, 42]	$(CH_2)_5CO_2H$	**1p** [14]	R' = H, CH$_3$CO; R^1 = nothing or (CH$_2$)$_2$O.
1h [41]	$(CH_2)_4O^tBu$	**1q** [14]	
1i [41, 42]	$(CH_2)_6SH$	**1r** [41, 49]	

Table 2.1 (continued)

No./Lit.	R	No./Lit.	R
1s [8]	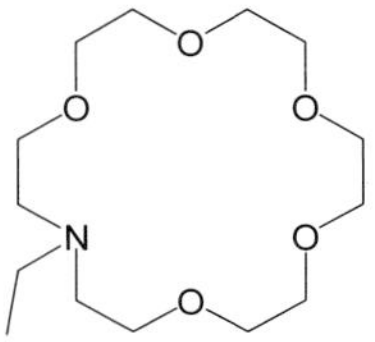	1w [48]	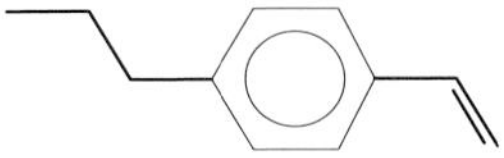
1t [9]	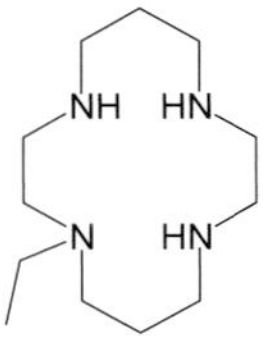	1x [48]	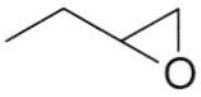
1u [50]	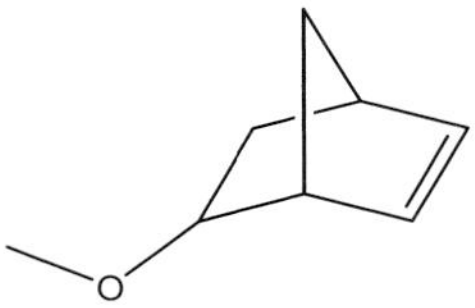	1y	$(CH_2)_4NCO$
1v [48]			

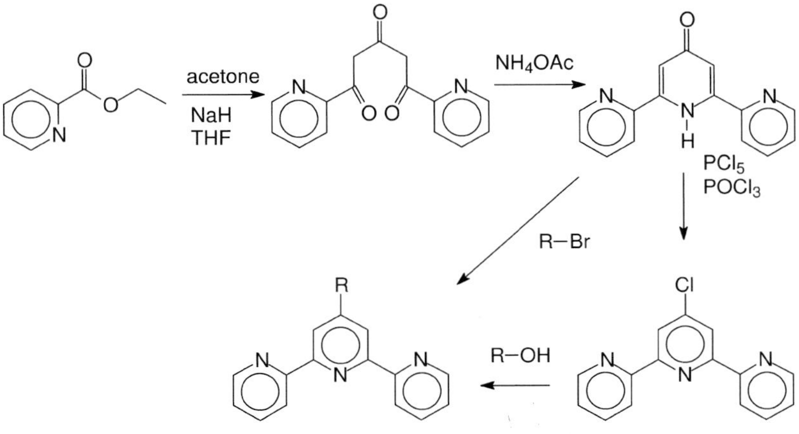

Scheme 2.7 Synthetic approach to terpyridinoxy derivatives via the chloroterpyridine and pyridine routes.

The reaction of 4-hydroxy-2,2,6,6-tetramethylpiperidin-1-oxyl (HO-TEMPO) with 4′-chloroterpyridine afforded the 4′-O-TEMPO-derivative **1m**, which represents a convenient spin-labeled terpyridine [45].

By treatment of 4-terpyridone with K_2CO_3 in DMF, followed by the addition of functionalized bromides and tosylates, various 4′-terpyridinoxy derivatives were obtained in high yields [48]. Among the examples are 10-bromodecyloxy, allyloxy, oxiranylmethoxy, 1-cyanopropyloxy, 4-vinylbenzyloxy, 2-[1-methoxyethoxy]ethoxy, and 2,7-*bis*[2-(2-oxyethoxy)ethoxy]naphthalene [51] groups.

Constable et al. [46, 47] reported the reaction of terpyridin-4′(1′-*H*)-one with 3-bromoprop-1-yne to give (56%) the alkyne-functionalized terpyridine ligand **1n**, which was subsequently treated with $B_{10}H_{14}$ in acetonitrile; however, only poor yields of the desired carbaborane-derivative **1o** (n = 1) were reported [47]. An analogous carbaborane **1o** (n = 3) was, however, prepared by treatment of 4′-hydroxyterpyridine with 1-(3-iodopropyl)-*closo*-1,2-carbaborane in the presence of potassium carbonate. Lithiated carbaborane cages (protection of the second CH group with SiMe$_2$^{t}Bu to prevent *bis*-lithiation) could also be reacted with chloroterpyridine, resulting in a directly linked carbaborane in an improved yield (36% over 2 reaction steps). In order to probe molecular recognition events by functionalization of biomolecules with metal-binding sites, Constable and Mundwiler [14] have also presented a new class of terpyridines possessing a sugar moiety. Glucosides have been attached directly or with a spacer-linkage (ethylene glycol) to 4′-hydroxyterpyridine by the use of α-bromo- or α-bromoethyl-glucose and their tetraacetyl-protected derivatives. The sugar-functionalized terpyridines **1p** were isolated in 27% (directly linked) and 68% yield (linked via ethylene glycol spacer), respectively [14]. Furthermore, a protected galactose derivative attached to the terpyridine unit at the 6-position of the sugar afforded (52%) **1q**.

Schubert et al. [49] investigated the special electronic properties associated with the novel ether-coupled examples, such as the fullerene-functionalized 4′-terpyridine **1r**, which was prepared (47%) by the reaction of **1k** with a chlorocarbonylfullerene. Fullerenes play an important role in the development of organic photophysical devices; thus, complex ligands such as **1r** could find applications as novel donor-acceptor arrays in organic solar cells.

Attachment of aza-crown macrocycles to the 4′-position of terpyridines was postulated to have important uses as luminescent or electrochemical sensors [52] and as di- or multi-topic terpyridine ligands [8, 9]. Thus, Ward et al. [8] described the preparation of 4′-substituted and 4′-(phenyl)-substituted terpyridines (see Section 2.2) with aza-18-crown-6-groups (**1s**) that were prepared (55%) by treatment of 4′-bromoterpyridine with aza-18-crown-6. Also, Martínez-Máñez et al. [9] reported a similar system by the reaction of 4′-(bromomethyl)terpyridine (see Section 2.3.3, **3bb**) with cyclam, which generated (50%) the functionalized 1,4,8,11-tetraazacyclotetradecane derivative **1t**.

The 4′-terpyridinoxynorbornene (**1u**) was prepared (61%) from 4′-chloroterpyridine and 5-norbornene-2-methanol via the Williamson ether synthesis [50]. Utilizing the nucleophilic substitution of the alcoholate of 4′-hydroxyterpyridine and functionalized bromides/tosylates ("pyridone-route"), other polymerizable groups were easily introduced [48], such as allyl (**1v**) and vinylbenzyloxy (**1w**) groups. Moreover, epoxide (**1x**) could be successfully connected to a terpyridine. The amine **1k** (see above) could be converted into an isocyanate group using di-*tert*-butyltricarbonate ("tri-carb"). The resulting **1y** represents a powerful building block for further functionalization, especially regarding supramolecular polymers.

By reacting linear aliphatic diols and thiols with chloroterpyridine, *bis*-terpyridines, separated by an alkyl spacer, were obtained [53].

A new approach to terpyridinoxy moieties has taken advantage of the Mitsunobu reaction; for example, terpyridone was reacted with alcohols in the presence of triphenylphosphine and diisopropylazodicarboxylate [54]. Because this methodology is very mild (ca. 2–3 hours at 25 °C), even sensitive or complicated functional groups, such as alkynes or nucleosides, can be introduced.

2.3.2
4′-Aryl-Substituted 2,2′:6′,2″-Terpyridines

4′-Arylterpyridines can be easily prepared by central ring assembly. One example is a modified Kröhnke reaction by treating 2-acetylpyridine with benzaldehyde and NaH (Scheme 2.8 and Table 2.2) [55, 56].

A series of 2,5-disubstituted 4′-arylterpyridines was prepared by Colbran et al. [57, 58] starting from 4′-(2,5-dimethoxyphenyl)terpyridine (**2a**, see also [59]), which was deprotected with hydrobromic acid to afford the hydroquinonyl ligand. The following conversions were conducted on *bis*[(2,5-dihydroxyphenyl)terpyridine]Ru(PF$_6$)$_2$, which, when treated with either benzoyl chloride, propanoyl chloride, or benzyl chloride, generated the corresponding benzoyl ester, propanoyl ester or benzyl ether derivatives.

Lo et al. [60] described the formation of a 4′-(4-aminophenyl)terpyridine (**2b**), via reduction [61] of the corresponding NO$_2$ substituent by treatment with hydrazine monohydrate and palladium on charcoal (see also [60, 62]). In the [4′-(4-aminophenyl)terpyridine]Ir(III) (PF$_6$)$_3$ complex, the amino-function was transformed into an isothiocyanate group (**2c**) by the reaction with CSCl$_2$ in presence of CaCO$_3$ in acetone. 4′-[4-(2,3,4,5-Tetramethylcyclopenta-1,3-dien-1-yl)phenyl]terpyridine was prepared by Siemeling et al. [63] in order to generate the corresponding ferrocene **2d** by lithiation followed by treatment with FeCl$_2$. The formation of boronic acid-functionalized 4′-arylterpyridines, e.g., **2e**, was

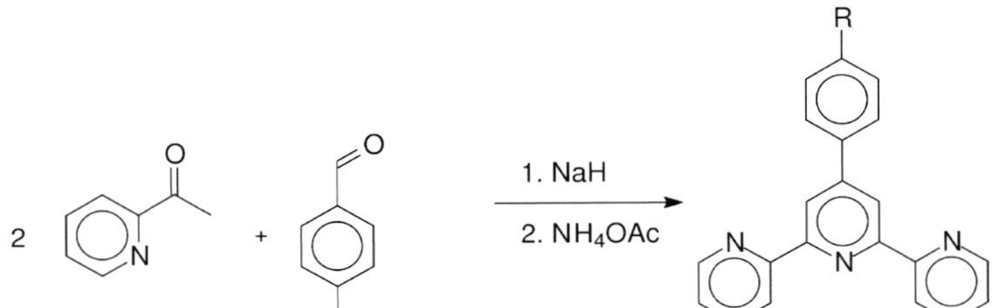

Scheme 2.8 Synthesis of aryl-substituted terpyridines.

Table 2.2 4′-Aryl substituted terpyridines.

No./Lit.	R	No./Lit.	R
2a [57]	R'O—⟨ring⟩—OR' (Me substituted); R' = CH$_3$, H, C(O)Ph, C(O)Et, CH$_2$Ph	2j [68]	PO(OEt)$_2$-*p*
2b [60]	C$_6$H$_4$-NH$_2$-*p*	2k [68]	PO$_3$H-*p*
2c [60]	C$_6$H$_4$-NCS-*p*	2l [69]	Me—⟨ring⟩(Br, Br)
2d [63]	—*p*-Ph— / —*p*-Ph— (ferrocene, Fe)	2m [69]	Me—⟨ring⟩(CN, CN)
2e [64]	-*p*-Ph—B⟨pinacol⟩	2n [70]	CH$_2$OC(O)CH$_3$-*p*
2f [64]	-*p*-Ph—B⟨dioxaborinane⟩	2o [70]	CH$_2$OH-*p*
2g [64]	B(OH)$_2$	2p [46, 71]	≡—X-*p*; X = CH$_2$OH, H, TMS
2h [65]	F-*p*	2q [61, 72]	*p*-OH
2i [66, 67]	^{t}Bu-*p*	2r [72]	O(CH$_2$)$_n$CO$_2$R'-*p*; n = 3,5,7,10; R' = Et, H

Table 2.2 (continued)

No./Lit.	R	No./Lit.	R
2s [72]	O(CH$_2$)$_5$CO$_2$R'- *m/o* R' = Et, H	2w [73]	
2t [75–77]		2x [74]	
2u [8]		2y [74]	
2v [9]		2z [78]	R = F, OR, OH

carried out by Aspley and Williams [64], in which *bis*(pinacolato)diboron was reacted with 4'-(4-bromophenyl)terpyridine. The analogous neopentyl ester **2f** was obtained by treatment of bromophenylterpyridine with *bis*(neopentyl-glyco-lato)diboron (as shown in Section 2.3.3 for ligand **3w**); **2f** was hydrolyzed to the boronic acid derivative **2g**.

4'-(4-Fluorophenyl)terpyridine (**2h**) was prepared by a two-step ring closure with 2-acetylpyridine and 4-fluorobenzaldehyde as the starting materials (step 1: 29%, step 2: 64%) [65]. 4'-(4-*Tert*-butylphenyl)terpyridine (**2i**), prepared by Constable et al. as an *oligo*pyridine with enhanced solubility properties [66, 67], was synthesized by the same method as that described for **2h**, starting with 2-acetylpyridine and 4-*tert*-butylbenzaldehyde (step 1: 40%, step 2: 73%).

Jing et al. [68] reported on the synthesis (87%) of 4-(terpyridin-4'-ylphenyl)-phosphonic acid (**2k**) by the saponification of the 4'-[*p*-PO(OEt)$_2$-phenyl]terpyridine (**2j**) with HCl (see also [79]). The 4'-(2-methyl-4,5-dicyanophenyl)terpyridine (**2m**) was synthesized in a seven-step synthesis [69] via the bromo-functionalized terpyridine precursor **2l**, followed by the Rosenmund-von Braun reaction.

4'-[4-(Bromomethyl)phenyl]terpyridine was transformed into the hydroxymethyl analog **2o** [70]; since direct substitution of the bromo-group was impractical, an indirect pathway via initial acetolysis with acetic acid and sodium acetate generated (85%) 4'-[4-(acetoxymethyl)phenyl]terpyridine (**2n**), which was subsequently saponified (65%) to form the desired product **2o** [70]. 4'-Arylterpyridines, functionalized in the para position with different alkyne substituents, were described by Constable et al. [46], in which different alkynes [HC≡CCH$_2$OH, HC≡CH, HC≡C(TMS)] were coupled with 4'-(4-bromophenyl)terpyridine in the presence of [PdCl$_2$(PPh$_3$)$_2$], CuI, and NEt$_3$ to afford (52–78%) **2p** [46, 71]. The angular monomer, 1,2-*bis*(terpyridin-4-ylethynyl)benzene, was prepared (41%) by the reaction of 1,2-diethynylbenzene with 4'-(trifluoromethanesulfonyloxy)terpyridine via Pd-catalyzed cross-coupling conditions using [C$_6$H$_5$)$_4$P]Pd(0) in basic solvent [80].

Ether substituents have also played an important role for 4'-aryl-substituted terpyridines because of their easy accessibility via substitution and condensation procedures. Hanabusa et al. [72] reported the synthesis of a series of 4'-(4-carboxyphenyl)terpyridines, in which 4'-[*p*-, *m*-, and o-carboxyphenyl-pentyloxy]terpyridines (**2r-s**) were obtained from the corresponding 4-, 3-, and 2-hydroxy-derivatives **2q** [61, 72] with 6-bromohexanoate (and homologs) or by using a modification of the Kröhnke method. Pikramenou et al. [75–77] investigated the creation of long-lived charge-separated states by the attachment of cyclodextrin receptors onto terpyridine ligands that can coordinate metal ions. Protection of all but one hydroxyl group of the β-cyclodextrin cups by methylation and subsequent reaction with 4'-[4-(bromomethyl)phenyl]terpyridine in THF in the presence of NaH yielded (71%) the desired 4'-cyclodextrin-functionalized terpyridine **2t**.

Similar to the synthesis described in Section 2.3.1, macrocycles (**1s-t**) were also attached to 4'-aryl-substituted terpyridines; for example, Ward et al. [8] described the preparation of 4'-phenyl-substituted terpyridines with aza-18-crown-6 groups **2u**, and Martínez-Máñez et al. [9] reported similar systems functionalized with 1,4,8,11-tetraazacyclotetradecane **2v**. Complexation of Cu(II) ions into the cavity of cyclam resulted in a quenching of the luminescence of the Ru(II) complex of the terpyridine moiety, making this system interesting as a "sensor" for heavy metal ions.

The 4'-[*p*-(1,4,7-triazacyclonon-1-ylmethyl)phenyl]terpyridine (**2w**) has been prepared (83%) by Moore et al. [73] by conversion of 4'-[4-(bromomethyl)phenyl]terpyridine with a "capped" 1,4,7-triazacyclononane in THF.

Jones et al. [74] prepared dibromothiophene-functionalized terpyridine luminescent receptor sites; for example, 4'-[4-(2,5-dibromothiophen-3-yl-methoxymethyl)phenyl]terpyridine (**2x**) was synthesized (60%) from (2,5-dibromothiophen-3-yl)methanol and 4'-[4-(bromomethyl)phenyl]terpyridine in the presence of NaH. 4'-[4-[2-(2,5-Dibromothiophen-3-yl)vinyl]phenyl]terpyridine (**2y**) was also prepared (80%) from the same bromomethyl derivative with triethylphosphite and 2,5-dibromothiophene-3-carbaldehyde.

A Kröhnke reaction with pentafluorobenzaldehyde led to the corresponding pentafluorinated phenylterpyridine **2z** [78]. The *para*-fluoro derivative could be replaced by alkoxy-groups and subsequently hydrolyzed to the alcohol, which could be used as precursor for further functionalization.

2.3.3
Other 4'-Functionalized 2,2':6''2'-Terpyridines

Phosphine-functionalized terpyridine entities are known to complex many different transition metal ions and to bind strongly to certain semiconductors [81]. The simplest example of this series is the 4'-(diphenylphosphino)terpyridine (**3a**), which was obtained by the treatment of 4'-chloroterpyridine with Li(PPh$_2$) in THF (Table 2.3) [82].

Diethyl terpyridine-4'-phosphonate (**3b**, see also [83]) was prepared from 4'-bromoterpyridine by treatment with HPO$_3$Et$_2$ in the presence of Pd(PPh$_3$)$_4$ [84, 85]. The 4'-CH$_2$P(O)Ph$_2$-functionalized terpyridine **3c** was synthesized from 4'-methyl-terpyridine with LDA and PPh$_2$Cl, then oxidized using NaIO$_4$. A double Wittig-Horner coupling then led to carotene-substituted terpyridine ligands [85]. 4'-[*Bis*(di-phenylphosphanylmethyl)]terpyridine (**3e**) was obtained by treatment of **3d** (pre-pared by the same method from 4'-methylterpyridine) with LDA and Ph$_2$PCl [86].

4'-(Phthalimidopropylsulfanyl)terpyridine (**3f**) and the 4'-[2-(1,3-dioxolan-2-yl)ethylsulfonyl] compound **3g** were synthesized from acetylpyridine, CS$_2$, and the corresponding alkyl halides by Sampath et al. [40]. Subsequently, the cyclic protective groups were converted into several 4'-alkylsulfanyl derivatives with amino (**3h**), hydroxy (**3i**), halo (**3j**), and aldehyde (**3k**) groups. 4'-Alkyl-based functional groups (**3l-p**) were introduced by initial deprotonation of 4'-methyl-terpyridine, followed by treatment with the corresponding alkyl halide. Maskus and Abruña [87] demonstrated the synthesis of 5-(terpyridin-4'-yl)pentane-1-thiol (**3r**) via consecutive treatment of the corresponding 5-chloropentyl derivative **3q**, obtained from 4'-methylterpyridine after deprotonation and conversion with 1-bromo-4-chlorobutane and NaOH, then dilute sulfuric acid.

Padilla-Tosta et al. [88] reported the preparation of 4'-(2-ferrocenyl-2-hydroxy-ethyl)terpyridine (**3s**), which was obtained (80%) by addition of ferrocene-carbaldehyde to 4'-methylterpyridine in the presence of LDA. Dehydration of the product yielded (25%) the corresponding 4'-(ferrocenylvinyl)terpyridine (**3t**, see also [97–99]).

Khatyr and Ziessel [89] reported **3u**, possessing terpyridines substituted with L-tyrosine fragments, from the optically active L-tyrosyl moieties possessing an iodo-functionality, which can be cross-coupled with 4'-ethynylterpyridine [100] in the presence of Pd(PPh$_3$)$_2$Cl$_2$ (R = H: 42%, R = COPh: 49%). 4'-Ethynylterpyridines are very versatile building blocks that can be readily synthesized by cross-coupling of alkynes with 4'-bromoterpyridines (Scheme 2.9). Treatment of 4'-(trifluoro-methylsulfonyl)oxyterpyridine [101] with phenylacetylene in the presence of Pd(PPh$_3$)$_4$ in THF containing diisopropylamine at 95 °C for 16 h gave (74%) the desired 4'-phenylethynylterpyridine [102]. The 4'-monoethynyl-*bis*terpyridine-Ru(II) complex has been efficiently coupled with diverse aryl iodides to generate extended systems [103].

Table 2.3 Other 4′-functionalized 2,2′:6′,2″-terpyridines.

No./Lit.	R	No./Lit.	R
3a [82]	PPh$_2$	**3k** [40]	S(CH$_2$)$_2$CHO
3b [84]	PO$_3$Et$_2$	**3l** [40]	(CH$_2$)$_3$CHO
3c [85]	CH$_2$P(O)Ph$_2$	**3m** [40]	(CH$_2$)$_3$CH$_2$OH
3d [85]	CH$_2$PPh$_2$	**3n** [40]	(CH$_2$)$_3$CH$_2$Br
3e [85]	CH(PPh$_2$)$_2$	**3o** [40]	S—(CH$_2$)$_4$—N
3f [40]	S—(CH$_2$)$_3$—N	**3p** [40]	(CH$_2$)$_3$CH$_2$NH$_2$
3g [40]	S—(CH$_2$)$_3$	**3q** [87]	(CH$_2$)$_5$Cl
3h [40]	S(CH$_2$)$_3$NH$_2$	**3r** [87]	(CH$_2$)$_5$SH
3i [40]	S(CH$_2$)$_3$OH	**3s** [88]	
3j [40]	S(CH$_2$)$_2$Cl	**3t** [88]	

Table 2.3 (continued)

No./Lit.	R	No./Lit.	R
3u [89]	L-tyrosine-fragments, R' = H, COPh.	**3bb** [9, 88]	CH_2Br
3v [64]		**3cc** [90]	
3w [91, 92]	$CO_2C_4H_9$	**3dd** [90]	
3x [62]	NO_2	**3ee** [90, 93]	
3y [62]	NH_2	**3ff** [90]	
3z [94]	N_3	**3gg** [95]	1. X = O; R = R' = H 2. X = O; R = NO_2; R' = H 3. X = O; R = Br; R' = H 4. X = NH; R = R' = H 5. X = S; R = Br; R' = H 6. X = S; R = Br; R' = Br 7. X = S; R = piperidine; R1 = Br
3aa [9]	CH_2OH	**3hh** [96]	$SnBu_3$

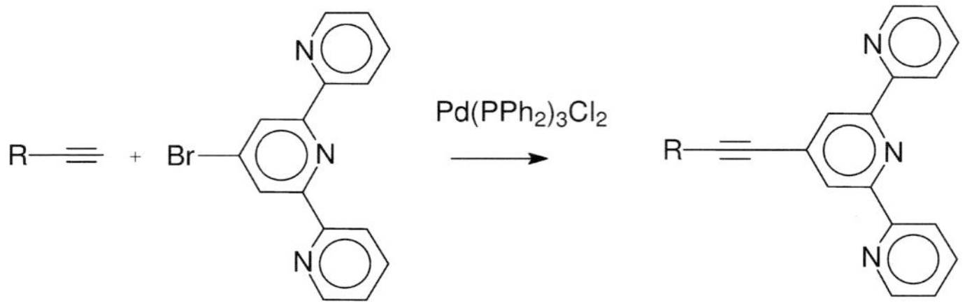

Scheme 2.9 Pd-catalyzed cross-coupling to ethynyl-terpyridines.

The use of these alkyne couplings is straightforward and opens up interesting avenues to various polytopic terpyridine ligands that can be used in the construction of supramolecular assemblies. Because of the extended conjugation systems, these compounds are characterized by outstanding optical properties such as room temperature luminescence of the corresponding ruthenium complexes. Furthermore, they have notable potential as "nanowires." A large variety of these compounds have been prepared in the laboratory of Ziessel, who reviewed this topic in 1999 [104]. This coupling procedure is very versatile and allows the construction of diverse *mono-* and *bis*-terpyridines containing extended conjugated alkyne units.

Aspley and Williams [64] synthesized 4'-boronate ester-substituted terpyridine ligands utilizing the Pd-catalyzed Miyaura [29] cross-coupling reaction; thus, ligand **3v** was prepared (69%) from 4'-bromoterpyridine, *bis*(neopentylglycolato)diboron, KOAc, and [1,1'-*bis*(diphenylphosphino)ferrocene]PdCl$_2$ [Pd(dppf)$_2$Cl$_2$]. Butyl terpyridine-4'-carboxylate (**3w**) was obtained (76%) by Pd-catalyzed carboalkoxylation of 4'-[(trifluoromethylsulfonyl)oxy]terpyridine [101] with CO in *n*-BuOH and Bu$_3$N [91, 92].

Fallahpour et al. [62] conveniently introduced nitro and amino groups into the 4'-position of the terpyridine ligands via the Stille reaction; for example, 2,6-dibromo-4-nitropyridine, which was obtained from 2,6-dibromopyridine-*N*-oxide, was coupled with 2-tributylstannylpyridine to afford (68%) 4'-nitroterpyridine (**3x**). Reduction of **3x** with hydrazine hydrate in the presence of palladium on charcoal resulted (76%) in the formation of 4'-aminoterpyridine (**3y**); whereas, 4'-azido-terpyridine (**3z**) was obtained (70%) by conversion of the 4'-nitroterpyridine (**3x**) with NaN$_3$ in DMF [94].

4'-(Hydroxymethyl)terpyridine (**3aa**) was synthesized by Martínez-Máñez et al. [9, 88] by the reduction of 4'-formylterpyridine [16] with NaBH$_4$ in THF in 85% yield. Subsequent reaction of **3aa** with CBr$_4$ and triphenylphosphane in CH$_2$Cl$_2$ yielded (40%) 4'-(bromomethyl)terpyridine (**3bb**).

2-Furyl- (**3cc**), 3-furyl- (**3dd**), 2-thienyl- (**3ee**) and 3-thienyl- (**3ff**) moieties were attached to the 4'-position of 2,2':6',2"- as well as other terpyridine structures [90] by treatment of 2- or 3-furaldehyde, or 2- or 3-formylthiophene, respectively, with 2-acetylpyridine in the presence of KOH in a methanol/water mixture. The obtained intermediates were converted (57–78%) to the corresponding terpyridines utilizing a modified Kröhnke synthesis. Constable et al. [93] reported the one-pot

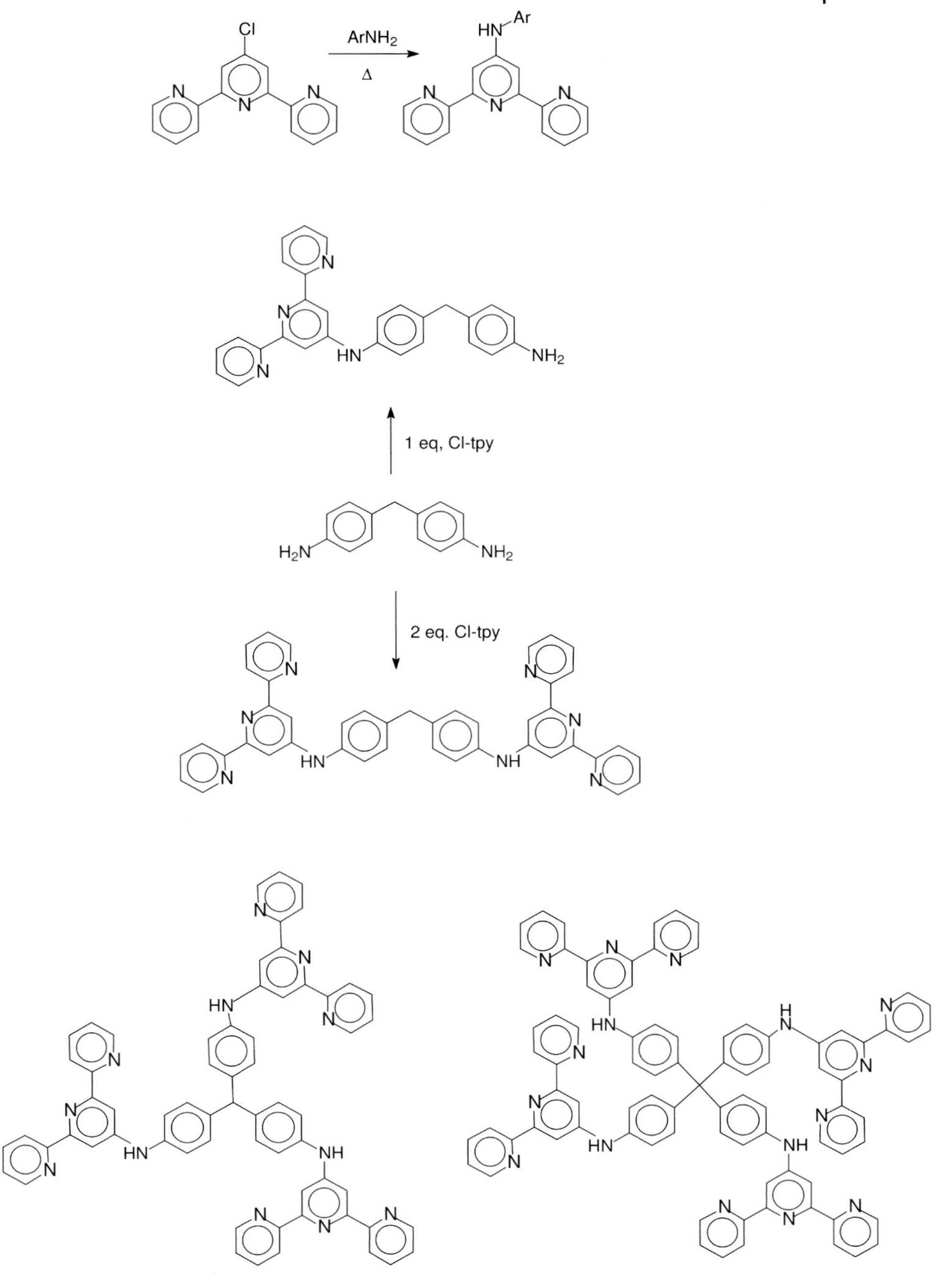

Scheme 2.10 Synthesis of multitopic terpyridines by reaction of amines with 4'-chloroterpyridine.

ring closure of two equivalents of 2-acetylpyridine and 2-thiophenecarbaldehyde via the intermediate diketone affording (40%) 4'-(2-thienyl)terpyridine (**3ee**). A relatively new contribution describing the partially solventless synthesis of terpyridines, functionalized with furanes, thiophenes, and pyrroles (**3gg**), has been described by Cave et al. [24–26], in which the first step of the Kröhnke reaction was performed solventlessly using Al_2O_3, and the second one was executed in the traditional fashion [95].

The thermally induced [4+2] cycloaddition of 3,5-*bis*(2-pyridyl)-1,2,4-triazine and tributyl(ethynyl)tin allowed the regioselective introduction of a tributylstannyl group yielding (73%) 4'-(tributylstannyl)terpyridine (**3hh**) [96]. As described in Sections 2.3.4 and 2.3.5, this method also enables the tributylstannyl functionalization of the terminal rings.

By reaction of 4'-chloroterpyridine with aromatic amines, various new terpyridines, including a tetra*kis*terpyridine, could be obtained (Scheme 2.10) by a nucleophilic aromatic substitution, in a manner comparable to the preparation of ethers [105]. No solvent was required, and the reaction was performed by simply heating the components.

Table 2.4 Terminally substituted 2,2':6',2''-terpyridines (unsymmetrical).

No./Lit.	R^1	R^2
4a [106]	6-Br	6''-CH_3
4b [106]	R + S	6''-CH_3
4c [108]	R + S	6''-H
4d [109]	6-Ph	6''-H
4e [109]	6-Ph + 4-Ph	6''-H
4f [96]	4-$SnBu_3$	4''-H
4g [96]	4-Br or 4-I	4''-H

2.3.4
Unsymmetrically Terminally Substituted 2,2':6',2"-Terpyridines

A pair of enantiomeric 6-norbornenyl-6"-methylterpyridines (**4b**) was prepared (58%) by Constable et al. [106] via the reaction of 6-bromo-6"-methylterpyridine (**4a**) synthesized via Kröhnke methodology with the sodium salts of (1*S*)-(–) or (1*R*)-(+)-borneol (Table 2.4). A similar method was used to prepare the chiral monosubstituted 6-(norbornenyl)terpyridines (**4c**) from 6-bromoterpyridine [107, 108]. The same group reported on the preparation of 6-phenyl- (**4d**; 66%) and 4,6-diphenyl-(**4e**; 71%) terpyridines utilizing Kröhnke methodology [109].

By the use of the appropriate tailor-made 1,2,4-triazines, Sauer et al. [96] synthesized 4-(tributylstannyl)terpyridine (**4f**) in a [4+2] cycloaddition with tributyl(ethynyl)tin in 54% yield. Subsequently, the tributylstannyl group could be converted into a bromo or iodo group with Br_2 or I_2 at –60 °C yielding (69 and 61%) the corresponding terpyridines **4g**, respectively.

2.3.5
Symmetrically Terminally Substituted 2,2':6',2"-Terpyridines

The synthesis of 6,6"-dimethylterpyridine [18] (**5a**) was improved (43%) by using a Stille-type cross-coupling of 2,6-dibromopyridine and 2-tributylstannyl-6-methylpyridine (Table 2.5) [35].

Constable et al. [109] used a Kröhnke ring-closure with different Mannich salts to yield (63%) **5a**. These authors also utilized the same method for the preparation of 6,6"-diphenylterpyridine (**5b**) with a Mannich salt and *N*-phenacylpyridinium bromide in 56% yield. Ring-closure of the *bis*-chalcone of 2,6-diacetylpyridine and *N*-phenacyl- or methyl-pyridinium bromide resulted in the analogous 4,4"-substituted **5c** (71%) and **5d** (61%). El-Ghayoury and Ziessel [91, 92] described the preparation of 6,6"-di-(*n*-butoxycarbonyl)terpyridine (**5e**) from 6,6"-dibromo-terpyridine, CO, and $Pd(PPh_3)_2Cl_2$ in *n*-BuOH and *n*-Bu_3N in 50% yield. Reduction of **5e** with $NaBH_4$ afforded (88%) 6,6"-*bis*-(hydroxymethyl)terpyridine (**5f**). Subsequent oxidation of **5f** with oxalyl chloride and DMSO in dichloromethane yielded (86%) 6,6"-diformylterpyridine (**5g**). Benniston [110] published the preparation of 6,6"-*bis*[4-(hydroxymethyl)phenyl]terpyridine (**5h**) by ring formation.

Galaup, Picard et al. [114] introduced terminal functionality by means of the *bis*-*N,N"*-oxide, which afforded the 6,6"-dinitrile (**5s**) under Reissert-Henze reaction conditions [115]. Subsequent conversion of **5s** to the corresponding acid (**5t**), ester (**5u**), or hydroxymethyl group (**5f**) to the desired bromomethyl (**5v**) functionality followed a traditional pattern. Displacement of the labile bromo groups with a protected triamine created an 18-membered hexaazamacrocycle incorporating the 2,2':6',2"-terpyridine moiety.

Sauvage et al. [39] described the Stille cross-coupling to improve (67%) the synthesis of 5,5"-dimethylterpyridine (**5i**) [116] by coupling 2-trimethylstannyl-5-methylpyridine and 2-bromo-5-methylpyridine. Schubert et al. [112] utilized 2-tri-butylstannyl-5-methylpyridine and 2-bromo-5-methylpyridine to enhance the yield

Table 2.5 Terminally substituted 2,2':6,2"-terpyridines (symmetric).

No.	R	No.	R
5a [35, 109]	6,6"-CH$_3$	5l [110, 111]	3,3"-p-C$_6$H$_4$-CH$_3$
5b [109]	6,6"-Ph	5m [113]	4,4"-Ph/6,6"-Ph-**R'**-*p* **R'** = COOH, COOEt, H, OMe
5c [109]	4,4"-Ph/6,6"-Ph	5n [22]	5,5"-CH=CHN(CH$_3$)$_2$
5d [109]	4,4"-CH$_3$/6,6"-Ph	5o [22]	5,5"-CHO
5e [91, 92]	6,6"-CO$_2$C$_4$H$_9$	5p [22]	4,4"-**R'**/5,5"-CH$_3$ **R'** = Ph, CHO, CH=CHN(CH$_3$)$_2$, CH$_2$OH, CO$_2$CH$_3$, CH$_3$, H.
5f [91, 92]	6,6"-CH$_2$OH	5q [96]	4,4"-SnBu$_3$
5g [91, 92]	6,6"-CHO	5r [96]	4,4"-Br; 4,4"-I
5h [110]	6,6"-p-C$_6$H$_4$-CH$_2$OH	5s [114]	6,6"-CN
5i [39, 112]	5,5"-Me	5t [114]	6,6"-COOH
5j [112]	5,5"-CH$_2$Br	5u [114]	6,6"-COOMe
5k [35]	4,4"-CH$_3$	5v [114]	6,6"-CH$_2$Br

(90%) of the same product. Subsequent bromination with NBS and AIBN in CCl$_4$ gave (30%) 5,5"-di(bromomethyl)terpyridine (**5j**). Schubert et al. [35] also described the synthesis of 4,4"-dimethylterpyridine (**5k**) via Stille-type cross coupling in 52% yield.

The uniquely hindered 3,3"-positions were also functionalized by Benniston et al. [110, 111] in order to introduce functionality from the "back" of the chelator; thus, 3,3"-di(4-*p*-tolyl)terpyridine (**5j**) was obtained via a 6-step synthesis including coupling and ring-closure reactions.

Mikel and Potvin [113] described the synthesis of different 4,4",6,6"-*tetra*-substituted terpyridines **5m**. For example, 4,4"-diphenyl-6,6"-di[4-(ethoxycarbonyl)phenyl]terpyridine was obtained (80%) from a double Kröhnke reaction; the 4,6,4",6"-tetraphenylterpyridine (98%) and 4,4"-diphenyl-6,6"-di-(4-methoxyphenyl)terpyridine (90%) were also prepared in a similar manner. Starting from 5,5"-dimethylterpyridine, Sasaki et al. [22] described the synthesis of 5,5"-di[(*N*,*N*'-dimethylamino)methylidene]terpyridine (**5n**) and 5,5"-diformylterpyridine (**5o**). The authors utilized *tert*-butoxy-*bis*(dimethylamino)methane (Bredereck's reagent) to form (47%) **5n** by heating 5,5"-dimethylterpyridine in DMF for several days. The formyl derivative **5o** was then obtained (51%) by oxidation of **5n** with NaIO$_4$.

By the use of 4,4'',5,5''-tetramethylterpyridine, which was obtained – as well as 4,4''-diphenyl-5,5''-dimethylterpyridine – via the Kröhnke procedure [2], the same authors synthesized the following terpyridines, summarized as **5p**: 4,4''-diformyl-5,5''-dimethylterpyridine by oxidation of 4,4'',5,5''-tetramethylterpyridine with H_2SeO_3 (42%) and 4,4''-*bis*[(N,N'-dimethylamino)methylidene]-5,5''-dimethylterpyridine by conversion with Bredereck's reagent in 67% yield. Furthermore, the 4,4''-diformylated terpyridine could be reduced with $NaBH_4$ in order to obtain the corresponding hydroxymethyl compound (77% yield).

As described in Section 2.3.4, Sauer et al. [96] prepared (47%) 4,4''-*bis*-(tributyl-stannyl)terpyridine (**5q**) from 2,6-*bis*(1,2,4-triazin-3-yl)pyridine and tributyl(ethy-nyl)tin. By subsequent treatment of **5q** with Br_2 or I_2, the corresponding bromo and iodo compounds **5r** in 79 and 53% yield, respectively, were formed.

2.3.6
Uniform All-Ring Substituted 2,2':6',2''-Terpyridines

A symmetrically functionalized terpyridine possessing one chloro moiety per ring (**6b**) was prepared by Cummings et al. [117] (Table 2.6).

Starting from 4,4',4''-trinitroterpyridine-N-oxide (**6a**) and excess of chlorine in glacial acetic acid, **6b** was obtained (70%), following a method described earlier by Case [118]. The 4,4',4''-, 4',5,5''-, and 4',6,6''-trimethylterpyridines **6c-e** (for **6c**, see also [119]) were also synthesized (50–60%) using Stille-type cross-coupling between the corresponding methyl-2-tributylstannylpyridines and 2,6-dibromo-

Table 2.6 Uniform all-ring substituted terpyridines.

No.	R
6a [117, 118]	4,4',4''-NO_2 – N-oxide
6b [117, 118]	4,4',4''-Cl
6c [33, 119]	4,4',4''-CH_3
6d [33]	4',5,5''-CH_3
6e [33]	4',6,6''-CH_3
6f [120]	4,4',4''-(5-nonyl)
6g [110]	4,4',4''-Ph
6h [121, 122]	4,4',4''-*tert*-butyl

4-methylpyridine [33]. 4,4′,4″-*Tris*(5-nonyl)terpyridine (**6f**) was obtained via the Pd-catalyzed coupling of 4-(5-nonyl)pyridine [120, 123]. Kröhnke condensation was similarly used to synthesize 4,4′,4″-*tris*phenylterpyridine (**6g**) [110].

2.3.7
Multifunctional 2,2′:6′,2″-Terpyridines with Variable Substituents

Terpyridines bearing different functional groups at the 4′-position and on the outer pyridine rings have recently expanded the versatility within the terpyridine family (Table 2.7).

Fallahpour et al. [23, 124] treated 4-ethoxy-2,6-diacetylpyridine with paraformaldehyde and dimethylamine in DMF to yield (71%) a Mannich salt as pink crystals, which with *N*-(methylacetyl)pyridinium chloride in ethanol readily resulted (56%) in the formation of 4′-ethoxy-6,6″-dimethylterpyridine (**7a**); subsequently, similar Kröhnke preparations gave **7b–7d**. The ethyl ether of **7b** was cleaved in pyridine and HCl to generate (50%) 4′-hydroxy-5,5″-dimethylterpyridine (**7e**).

Stille cross-coupling of 4-ethoxycarbonyl-2,6-dibromopyridine and 6-methyl-2-tributylstannylpyridine was used to synthesize (44%) 4′-ethoxycarbonyl-6,6″-dimethylterpyridine (**7f**) [34, 114]. Subsequently, the methyl groups were converted (27%) to bromomethyl groups (**7g**) via NBS bromination in benzene. In the same manner, the 4-carboxy- [35, 125] or (m)ethoxycarbonyl [35, 125]-5,5″-dimethyl-2,2′:6′,2″-terpyridines (**7h**) have been prepared in ca. 60% yield. Bromination of the ring methyl groups resulted in the formation of the corresponding 4′-(m)ethoxycarbonyl-5,5″-di(bromomethyl)terpyridine (**7i**) in 10 and 70% yields, respectively. Substitution of the bromo groups by acetate groups with NaOAc in acetic acid gave (70%) 4′-ethoxycarbonyl-5,5″-di(acetoxymethyl)terpyridine (**7j**) [35, 126]. Reduction of the 4′-ester group of **7h** with NaBH$_4$ in methanol yielded (85%) the 5,5″-dimethyl-4′-(hydroxymethyl)terpyridine (**7k**) [35, 126], which could be protected with TBDMS-Cl in pyridine to give (85%) ether **7l**.

Schlüter et al. [37] prepared various 4′-functionalized terpyridines with bromo or chloro groups at the terminal 5,5″-positions utilizing Stille cross-coupling conditions, for example, 2-trimethylstannyl-5-chloro- (or bromo-) pyridine and 2,6-dibromopyridines functionalized in the 4-position with CH$_2$OC$_6$H$_{13}$ or CH$_2$O-MEM (MEM = methoxyethoxymethoxy). These authors further synthesized the terpyridine **7m** (65%), **7n** (14%), and **7o** (26%). Fallahpour et al. [62] prepared 4′-nitro-5,5″-dimethylterpyridine (**7p**) utilizing the Stille conditions with 2,6-dibromo-4-nitropyridine and 5-methyl-2-tributylstannylpyridine in 64% yield. The nitro group was reduced (69%) to the amino moieties in the presence of palladium on charcoal and hydrazine hydrate to give **7q**. The reaction of the nitro derivative **7p** with NaN$_3$ in DMF afforded (72%) 4′-azido-5,5″-dimethylterpyridine (**7r**) [94]. Fallahpour et al. have also prepared the related monomethyl derivatives [62]. For example, 4′-nitro-5-methylterpyridine (**7s**) was obtained via Stille coupling of 2,6-dibromo-4-nitropyridine and 5-methyl-2-tributylstannylpyridine. The nitro group was converted to the corresponding amino function (**7t**) by reduction over Pd/C or to an azido group (**7u**) by treatment of the nitro derivative with NaN$_3$ in DMF

Table 2.7 Multifunctional terpyridines with variable substituents.

No.	R^1	R^2
7a [23, 124]	OCH$_2$CH$_3$	6,6"-CH$_3$
7b [23, 124]	OCH$_2$CH$_3$	5,5"-CH$_3$
7c [23, 124]	OCH$_2$CH$_3$	4,4"-CH$_3$
7d [23, 124]	OCH$_2$CH$_3$	6,6"-CH$_3$, 4,4"-p-toluene
7e [23, 124]	OH	5,5"-CH$_3$
7f [34]	CO$_2$CH$_2$CH$_3$	6,6"-CH$_3$
7g [34]	CO$_2$CH$_2$CH$_3$	6,6"-CH$_2$Br
7h [34, 35]	CO$_2$H or CO$_2$CH$_3$ or CO$_2$CH$_2$CH$_3$	5,5"-CH$_3$
7i [35, 126]	CO$_2$CH$_3$ or CO$_2$CH$_2$CH$_3$	5,5"-CH$_2$Br
7j [35, 126]	CO$_2$CH$_2$CH$_3$	5,5"-CH$_2$OAc
7k [35, 126]	CH$_2$OH	5,5"-CH$_3$
7l [35, 126]	CH$_2$OTBDMS	5,5"-CH$_3$
7m [37]	CH$_2$OC$_6$H$_{13}$	5,5"-Cl
7n [37]	CH$_2$OC$_6$H$_{13}$	5,5"-Br
7o [37]	CH$_2$O-MEM	5,5"-Br
7p [62]	NO$_2$	5,5"-CH$_3$
7q [62]	NH$_2$	5,5"-CH$_3$
7r [94]	N$_3$	5,5"-CH$_3$
7s [62]	NO$_2$	5-CH$_3$
7t [62]	NH$_2$	5-CH$_3$
7u [94]	N$_3$	5-CH$_3$
7v [91]	p-toluene	6,6"-COOC$_4$H$_9$
7w [114]	Me	6,6"-CH$_2$OH
7x [114]	Me	6,6"-CH$_2$OCH$_2$COOMe
7y [114]	Me(N-O)	6,6"-CH$_2$OCH$_2$COOMe
7z [114]	CH$_2$Br	6,6"-CH$_2$OCH$_2$COOMe

[94]. El-Ghayoury and Ziessel [91] et al. prepared (52%) 4′-(4-methylphenyl)-6,6″-di(*n*-butoxycarbonyl)terpyridine (**7v**) from the corresponding 6,6″-dibromo derivative [17] in the presence of Pd(PPh$_2$)Cl$_2$, CO, and Bu$_3$N in *n*-butanol.

Beley et al. [127] reported the five-step synthesis of 3′,4′-di(carboxy)terpyridine, shown to be an extremely powerful extractor of iron traces (ppm), from 2-acetyl-pyridine and furfural, when this acid was treated with POCl$_3$/PCl$_5$ and then methanol to afford the diester (28%) along with the interesting 4-chloro-3′,4′-di(methoxycarbonyl)terpyridine.

Galaup, Picard et al. [114] introduced terminal functionality as well as either a 4-ethoxycarbonyl or a -CH$_2$CH$_2$COOMe moiety via an initial Stille-type cross-coupling procedure with ethyl 2,6-dichloro-4-pyridinecarboxylate with 2-methyl-6-(tributylstannyl)pyridine [128] to generate the 4′-ethoxycarbonyl-6,6″-dimethyl-terpyridine (**7f**), which was reduced to the 4′-hydroxymethyl derivative (**7w**). Protection of the hydroxy moiety (**7w**) was conducted by treatment with methyl α-bromoacetate; subsequent conversion of **7y** to **7z** followed a traditional sequence: conversion to the *bis*-N-oxide via treatment with 3-chloroperbenzoic acid, then rearrangement by perfluoroacetic anhydride, and finally nucleophilic substitution with LiBr.

The novel synthesis of highly (symmetrically) substituted terpyridines via a triazine intermediate has recently appeared; initially 2,6-diformylpyridine is condensed with two equivalents of an isonitrosoacetophenone hydrazone to afford (52%) the 2,6-*bis*[6-(4-aryl)-1,2,4-triazin-3-yl-4-oxide]pyridine, which, on treatment with acetone cyanohydrin and triethylamine under reflux, generated (55%) the key intermediate 2,6-*bis*[5-cyano-6-(aryl)-1,2,4-triazin-3-yl]pyridine, whose conversion (90%) to 5,5″-*bis*(aryl)-6,6″-dicyanoterpyridine was accomplished via heating in the presence of bicycle[2.2.1]hepta-2,5-diene in toluene [129].

2.4
Summary and Outlook

The large number of interesting applications of functionalized 2,2′:6′,2″-ter-pyridines capable of metal ion coordination has expanded their versatility in the fields of supramolecular and macromolecular chemistry as well as electro-chemistry. Modern ring-assembly and cross-coupling procedures enable the well-directed introduction of different functionalities into almost every position of the terpyridine ring system. Nevertheless, the overall molecular diversity of these terpyridine derivatives is still comparatively small, since the methods for their construction are only compatible with a few less-reactive functional groups. Therefore, subsequent functional group conversions have to be carried out to enhance the "pool" of terpyridine ligands. Particularly in the field of multifunctional terpyridines, further progress is required to introduce different highly reactive groups into one molecule. This could allow the incorporation of the chelating ligands into more complex architectures and would open avenues to novel materials in the fields of supramolecular, polymer, or surface chemistry.

References

1 A. M. W. Cargill-Thompson, *Coord. Chem. Rev.* **1997**, *160*, 1–52.

2 R.-A. Fallahpour, *Synthesis* **2003**, 155–184.

3 F. Pezet, I. Sasaki, J. C. Daran, J. Hydrio, H. Ait-Haddou, G. Balavoine, *Eur. J. Inorg. Chem.* **2001**, 2669–2674.

4 H.-L. Kwong, W.-S. Lee, *Tetrahedron: Asymmetry* **2000**, *11*, 2299–2308.

5 H.-L. Kwong, W.-L. Wong, W.-S. Lee, L.-S. Cheng, W.-T. Wong, *Tetrahedron: Asymmetry* **2001**, *12*, 2683–2694.

6 G. Chelucci, A. Saba, F. Soccolini, D. Vignola, *J. Mol. Cat. A: Chemical.* **2002**, *178*, 27–33.

7 K. L. Bushell, S. M. Couchman, J. C. Jeffery, L. H. Rees, M. D. Ward, *J. Chem. Soc., Dalton Trans.* **1998**, 3397–3403.

8 B. Whittle, S. R. Batten, J. C. Jeffery, L. H. Rees, M. D. Ward, *J. Chem. Soc., Dalton Trans.* **1996**, 4249–4255.

9 M. E. Padilla-Tosta, J. M. Lloris, R. Martínez-Máñez, A. Benito, J. Soto, T. Pardo, M. A. Miranda, M. D. Marcos, *Eur. J. Inorg. Chem.* **2000**, 741–748.

10 B. Galland, D. Limosin, H. Laguitton-Pasquier, A. Deronzier, *Inorg. Chem. Commun.* **2002**, *5*, 5–8.

11 F. Loiseau, C. Di Pietro, S. Serroni, S. Campagna, A. Licciardello, A. Manfredi, G. Pozzi, S. Quici, *Inorg. Chem.* **2001**, *40*, 6901–6909.

12 P. Laine, F. Bedioui, P. Ochsenbein, V. Marvaud, M. Bonin, E. Amouyal, *J. Am. Chem. Soc.* **2002**, *124*, 1364–1377.

13 G. D. Storrier, S. B. Colbran, D. C. Craig, *J. Chem. Soc., Dalton Trans.* **1997**, 3011–3028.

14 E. C. Constable, S. Mundwiler, *Polyhedron* **1999**, *18*, 2433–2444.

15 F. Kröhnke, *Synthesis* **1976**, 1–24.

16 K. T. Potts, A. Usifer, H. D. Guadalupe, H. D. Abruña, *J. Am. Chem. Soc.* **1987**, *109*, 3961–3966.

17 E. C. Constable, J. Lewis, *Polyhedron* **1982**, *1*, 303–304.

18 G. R. Newkome, D. C. Hager, G. E. Kiefer, *J. Org. Chem.* **1986**, *51*, 850–853.

19 D. C. Owsley, J. M. Nelke, J. J. Bloomfield, *J. Org. Chem.* **1973**, *38*, 901–903.

20 D. L. Jameson, L. E. Guise, *Tetrahedron Lett.* **1991**, *32*, 1999–2002.

21 J. C. Adrian, L. Hassib, N. De Kimpe, M. Keppens, *Tetrahedron* **1998**, *54*, 2365–2370.

22 I. Sasaki, J. C. Daran, G. G. A. Balavoine, *Synthesis* **1999**, 815–820.

23 R.-A. Fallahpour, M. Neuburger, M. Zehnder, *Polyhedron* **1999**, *18*, 2445–2454.

24 G. W. V. Cave, C. L. Raston, *Chem. Commun.* **2000**, 2199–2200.

25 G. W. V. Cave, C. L. Raston, *J. Chem. Soc., Perkin Trans. 1* **2001**, 3258–3264.

26 G. W. V. Cave, C. L. Raston, J. L. Scotta, *Chem. Commun.* **2001**, 2159–2169.

27 J. Uenishi, T. Tanaka, S. Wakabayashi, S. Oae, *Tetrahedron Lett.* **1990**, 4625–4628.

28 J. E. Parks, B. E. Wagner, R. H. Holm, *J. Organomet. Chem.* **1973**, *56*, 53–66.

29 N. Miyaura, A. Suzuki, *Chem. Rev.* **1995**, *95*, 2457–2483.

30 E. Negishi, *Current Trends in Organic Synthesis*, Pergamon, New York, **1983**.

31 J. K. Stille, *Angew. Chem. Int. Ed. Engl.* **1986**, *25*, 508–523.

32 D. J. Cardenas, J.-P. Sauvage, *Synlett* **1996**, 916–918.

33 R.-A. Fallahpour, *Synthesis* **2000**, 1665–1667.

34 G. Ulrich, S. Bedel, C. Picard, P. Tisnes, *Tetrahedron Lett.* **2001**, *42*, 6113–6115.

35 M. Heller, U. S. Schubert, *Synlett* **2002**, 751–754.

36 U. S. Schubert, C. Eschbaumer, *Org. Lett.* **1999**, *1*, 1027–1029.

37 U. Lehmann, O. Henze, A. D. Schlüter, *Chem Eur. J.* **1999**, *5*, 854–859.

38 M. Chavarot, Z. Pikramenou, *Tetrahedron Lett.* **1999**, *40*, 6865–6868.

39 S. A. Savage, A. P. Smith, C. L. Fraser, *J. Org. Chem.* **1998**, *63*, 10048–10051.

40 U. Sampath, W. C. Putnam, T. A. Osiek, S. Touami, J. Xie, D. Cohen, A. Cagnolini, P. Droege, D. Klug, C. L. Barnes, A. Modak, J. K. Bashkin, S. S. Jurisson, *J. Chem. Soc., Dalton Trans.* **1999**, 2049–2058.

41 U. S. Schubert, C. Eschbaumer, O. Hien, P. R. Andres, *Tetrahedron Lett.* **2001**, *42*, 4705–4707.

42 U. S. Schubert, P. R. Andres, H. Hofmeier, *Polym. Mater.: Sci. Eng.* **2001**, *85*, 510–511.

43 G. R. Newkome, R. Güther, C. N. Moorefield, F. Cardullo, L. Echegoyen, E. Pérez-Cordero, H. Luftmann, *Angew. Chem. Int. Ed. Engl.* **1995**, *34*, 2023–2026.

44 X. Liu, E. J. L. McInnes, C. A. Kilner, M. Thornton-Pett, M. A. Halcrow, *Polyhedron* **2001**, *20*, 2889–2900.

45 M. A. Halcrow, E. K. Brechin, E. J. L. McInnes, F. E. Mabbs, J. E. Davies, *J. Chem. Soc., Dalton Trans.* **1998**, 2477–2482.

46 E. C. Constable, C. E. Housecroft, L. A. Johnston, D. Armspach, M. Neuburger, M. Zehnder, *Polyhedron* **2001**, *20*, 483–492.

47 D. Armspach, E. C. Constable, C. E. Housecroft, M. Neuburger, M. Zehnder, *J. Organomet. Chem.* **1997**, *550*, 193–206.

48 P. R. Andres, H. Hofmeier, B. G. G. Lohmeijer, U. S. Schubert, *Synthesis* **2003**, 2865–2871.

49 A. El-Ghayoury, A. P. H. J. Schenning, P. v. Hal, C. Weidl, J. v. Dongen, R. A. J. Janssen, U. S. Schubert, E. W. Meijer, *Thin Solid Films* **2002**, *403/404*, 97–101.

50 R. Kröll, C. Eschbaumer, U. S. Schubert, M. R. Buchmeiser, K. Wurst, *Macromol. Chem. Phys.* **2001**, *202*, 645–653.

51 H. S. Chow, E. C. Constable, C. E. Housecroft, M. Neuburger, *J. Chem. Soc., Dalton Trans.* **2004**, 4568–4569.

52 M. Schmittel, H. Ammon, *Chem. Commun.* **1995**, 687–688.

53 P. R. Andres, U. S. Schubert, *Synthesis* **2004**, 1229–1238.

54 J. Hovinen, *Tetrahedron Lett.* **2004**, *45*, 5707–5709.

55 E. C. Constable, J. Lewis, M. C. Liptrot, P. R. Raithby, *Inorg. Chim. Acta.* **1990**, *178*, 47–54.

56 E. C. Constable, A. M. W. C. Thompson, N. Armaroli, V. Balzani, M. Maestri, *Polyhedron* **1992**, *11*, 2707–2709.

57 G. D. Storrier, S. B. Colbran, D. C. Craig, *J. Chem. Soc., Dalton Trans.* **1998**, 1351–1363.

58 G. D. Storrier, S. B. Colbran, D. B. Hibbert, *Inorg. Chim. Act.* **1995**, *239*, 1–4.

59 C. A. Howard, M. D. Ward, *Angew. Chem. Int. Ed. Engl.* **1992**, *31*, 1028–1030.

60 K. K.-W. Lo, C.-K. Chung, D. C.-M. Ng, N. Zhu, *New J. Chem.* **2002**, *26*, 81–88.

61 W. Spahni, G. Calzaferri, *Helv. Chim. Acta* **1984**, *67*, 450–454.

62 R.-A. Fallahpour, M. Neuburger, M. Zender, *New J. Chem.* **1999**, *23*, 53–61.

63 U. Siemling, U. Vorfeld, B. Neumann, H.-G. Stammler, M. Fontani, P. Zanello, *J. Org. Met. Chem.* **2001**, 733–737.

64 C. J. Aspley, J. A. G. Williams, *New J. Chem.* **2001**, *25*, 1136–1147.

65 E. C. Constable, M. Neuburger, A. P. Smith, M. Zehnder, *Inorg. Chim. Acta* **1998**, *275–276*, 359–365.

66 E. C. Constable, P. Harverson, D. R. Smith, L. Whall, *Polyhedron* **1997**, *16*, 3615–3623.

67 E. C. Constable, P. Harverson, D. R. Smith, L. A. Whall, *Tetrahedron* **1994**, *50*, 7799–7806.

68 B. Jing, H. Zhang, M. Zhang, Z. Lu, T. Shen, *J. Mater. Chem.* **1998**, *8*, 2055–2060.

69 M. Kimura, T. Hamakawa, K. Hanabusa, H. Shirai, N. Kobayashi, *Inorg. Chem.* **2001**, *40*, 4775–4779.

70 M. Heller, U. S. Schubert, *e-polymers* **2002**, *27*, 1–11.

71 V. Grosshenny, R. Ziessel, *J. Organomet. Chem.* **1993**, *453*, C19–C22.

72 K. Hanabusa, T. Hirata, D. Inoue, M. Kimura, H. Shirai, *Coll. Surf. A: Physicochem. Eng.* **2000**, *169*, 307–315.

73 M. L. Turonek, P. Moore, W. Errington, *J. Chem. Soc., Dalton Trans.* **2000**, 441–444.

74 Y. Zhang, C. B. Murphy, W. E. Jones, *Macromolecules* **2002**, *35*, 630–636.

75 J. M. Haider, M. Chavarot, S. Weidner, I. Sadler, R. M. Williams, L. De Cola, Z. Pikramenou, *Inorg. Chem.* **2001**, *40*, 3912–3921.

76 J. M. Haider, Z. Pikramenou, *Eur. J. Inorg. Chem.* **2001**, 189–194.

77 S. Weidner, Z. Pikramenou, *Chem. Commun.* **1998**, 1473–1474.

78 E. C. Constable, B. Kariuki, A. Mahmood, *Polyhedron* **2003**, *22*, 687–698.

79 H. Toshikazu, M. Toshio, O. Yoshiki, A. Toshio, *Synthesis* **1981**, 56–58.

80 S.-H. Hwang, C. N. Moorefield, F. R. Fronczek, O. Lukoyanova, L. Echegoyen, G. R. Newkome, *Chem. Commun.* **2005**, 713–715.

81 R. Ziessel, *Synthesis* **1999**, 1839–1865.

82 E. C. Constable, C. E. Housecroft, M. Neuburger, A. G. Schneider, M. Zehnder, *J. Chem. Soc., Dalton Trans.* **1997**, 2427–2434.

83 P. Pechy, F. P. Rotzinger, M. K. Nazeeruddin, O. Kohle, S. M. Zakeeruddin, R. Humphry-Baker, M. Grätzel, *Chem. Commun.* **1995**, 65–66, 1093.

84 S. M. Zakeeruddin, M. K. Nazeeruddin, P. Pechy, F. P. Rotzinger, R. Humphry-Baker, K. Kalyanasundaram, M. Grätzel, *Inorg. Chem.* **1997**, *36*, 5937–5946.

85 G. Pickaert, R. Ziessel, *Tetrahedron Lett.* **1998**, *39*, 3497–3500.

86 G. Pickaert, M. Cesario, L. Douce, R. Ziessel, *Chem. Commun.* **2000**, 1125–1126.

87 M. Maskus, H. D. Abruña, *Langmuir* **1996**, *12*, 4455–4462.

88 M. E. Padilla-Tosta, R. Martínez-Máñez, J. Soto, J. M. Lloris, *Tetrahedron* **1998**, *54*, 12039–12046.

89 A. Khatyr, R. Ziessel, *Synthesis* **2001**, 1665–1670.

90 L.-X. Zhao, T. S. Kim, S.-H. Ahn, T.-H. Kim, E.-K. Kim, W.-J. Cho, H. Choi, C.-S. Lee, J.-A. Kim, T. C. Jeong, C.-J. Chang, E.-S. Lee, *Bioorg. Med. Chem. Lett.* **2001**, *11*, 2659–2662.

91 A. El-Ghayoury, R. Ziessel, *J. Org. Chem.* **2000**, *65*, 7757–7763.

92 A. El-Ghayoury, R. Ziessel, *Tetrahedron Lett.* **1998**, *39*, 4473–4476.

93 S. Encinas, L. Flamigni, F. Barigelletti, E. C. Constable, C. E. Housecroft, E. R. Schofield, E. Figgemeier, D. Fenske, M. Neuburger, J. G. Vos, M. Zehnder, *Chem. Eur. J.* **2002**, *8*, 137–150.

94 R.-A. Fallahpour, M. Neuburger, M. Zehnder, *Synthesis* **1999**, 1051–1055.

95 J. Husson, E. Migianu, M. Beley, G. Kirsch, *Synthesis* **2004**, 267–270.

96 J. Sauer, D. K. Heldmann, G. R. Pabst, *Eur. J. Org. Chem.* **1999**, 313–321.

97 B. König, M. Nimtz, H. Zieg, *Tetrahedron* **1995**, *51*, 6267–6272.

98 E. C. Constable, A. J. Edwards, R. Martínez-Máñez, P. R. Raithby, A. M. W. Cargill-Thompson, *J. Chem. Soc., Dalton Trans.* **1995**, 3253–3256.

99 S.-H. Hwang, C. N. Moorefield, F. R. Fronczek, O. Lukoyanova, L. Echegoyen, G. R. Newkome, *Chem. Commun.* **2005**, 713–715.

100 R. Ziessel, J. Suffert, M.-T. Youinou, *J. Org. Chem.* **1996**, *61*, 6535–6546.

101 K. T. Potts, D. Konwar, *J. Org. Chem.* **1991**, *56*, 4815–4816.

102 A. C. Benniston, G. Chapman, A. Harriman, M. Mehrabi, C. A. Sams, *Inorg. Chem.* **2004**, *43*, 4227–4233.

103 A. C. Benniston, S. Mitchell, S. A. Rostron, S. Yang, *Tetrahedron Lett.* **2004**, *45*, 7883–7885.

104 R. Ziessel, *Synthesis* **1999**, 1839–1865.

105 E. C. Constable, C. E. Housecroft, Y. Tao, *Synthesis* **2004**, 869–874.

106 E. C. Constable, T. Kulke, M. Neuburger, M. Zehnder, *Chem. Commun.* **1997**, 489–490.

107 R. Chotalia, E. C. Constable, M. J. Hannon, D. A. Tocher, *J. Am. Chem. Soc., Dalton Trans.* **1995**, 3571–3580.

108 E. C. Constable, C. E. Housecroft, T. Kulke, C. Lazzarini, E. R. Schofield, Y. Zimmermann, *J. Chem. Soc., Dalton Trans.* **2001**, 2864–2871.

109 E. C. Constable, G. Baum, E. Bill, R. Dyson, R. Eldik, D. Fenske, S. Kaderli, D. Morris, A. Neubrand, M. Neuburger, D. R. Smith, K. Wieghardt, M. Zehnder, A. D. Zuberbühler, *Chem. Eur. J.* **1999**, *5*, 498–508.

110 A. C. Benniston, *Tetrahedron Lett.* **1997**, *47*, 8279–8282.

111 A. C. Benniston, L. J. Farrugia, P. R. Mackie, P. Mallinson, W. Clegg, S. J. Teat, *Aust. J. Chem.* **2000**, *53*, 707–713.

112 U. S. Schubert, C. Eschbaumer, G. Hochwimmer, *Synthesis* **1999**, 779–782.

113 C. Mikel, P. G. Potvin, *Inorg. Chim. Act.* **2001**, *325*, 1–8.

114 C. Galaup, J.-M. Couchet, S. Bedel, P. Tisnès, C. Picard, *J. Org. Chem.* **2005**, *70*, 2274–2284.

115 W. K. Fife, *J. Org. Chem.* **1983**, *48*, 1375–1377.

116 A. Livoreil, C. O. Dietrich-Buchecker, J.-P. Sauvage, *J. Am. Chem. Soc.* **1994**, *116*, 9399–9400.

117 S. E. Hobert, J. T. Carney, S. D. Cummings, *Inorg. Chim. Acta* **2001**, *318*, 89–96.

118 F. H. Case, *J. Org. Chem.* **1962**, *27*, 640–641.

119 P. E. Rosevear, W. H. F. Sasse, *J. Heterocycl. Chem.* **1971**, *8*, 483–485.

120 G. Kickelbick, K. Matyjaszewski, *Macromol. Rapid Commun.* **1999**, *20*, 341–346.

121 S. B. Billings, M. T. Mock, K. Wiacek, M. B. Turner, W. S. Kassel, K. J. Takeuchi, A. L. Rheingold, W. J. Boyko, C. A. Bessel, *Inorg. Chim. Acta* **2003**, *355*, 103–115.

122 K. M. C. Wong, W.-S. Tang, B. W. K. Chu, N. Zhu, V. W. W. Yam, *Organometallics* **2004**, *23*, 3459–3465.

123 K. Matyjaszewski, T. E. Patten, J. Xia, *J. Am. Chem. Soc.* **1997**, *119*, 674–680.

124 R.-A. Fallahpour, E. C. Constable, *J. Chem. Soc., Perkin Trans. 1* **1997**, 2263–2264.

125 R.-A. Fallahpour, *Synthesis* **2000**, 1138–1142.

126 M. Heller, U. S. Schubert, *J. Org. Chem.* **2002**, *67*, 8269–8272.

127 M. Beley, C.-A. Bignozzi, G. Kirsch, M. Alebbi, J.-C. Raboin, *Inorg. Chim. Acta* **2000**, *318*, 197–200.

128 G. S. Hanan, U. S. Schubert, E. R. D. Volkmar, J.-M. Lehn, N. Kyritsakas, J. Fischer, *Can. J. Chem.* **1997**, *75*, 169–182.

129 V. N. Kozhevnikov, D. M. Kozhevnikov, V. L. Rusinov, O. N. Chapakhin, B. König, *Synthesis* **2003**, 2400–2404.

3
Chemistry and Properties of Terpyridine Metal Complexes[*]

3.1
Introduction

In this chapter, an overview of the latest achievements in the field of mononuclear terpyridine complexes is presented. The emphases are directed toward ruthenium(II) complexes and their optical properties with a special focus on their properties (e.g., redox properties, luminescence, etc.). Potential applications in optical nano-devices, molecular storage units, molecular switches [1], or solar cells are presented. The related, extended-supramolecular architectures, e.g., dyads, triads, and cyclic structures, will be presented in Chapter 4.

3.2
Synthetic Strategies

3.2.1
Metal Complexes

*Bis*terpyridine metal complexes of the type $[M(tpy)_2(X)_2]$ (X = e.g. Cl^-, ClO_4^-, PF_6^-) have been known for a long time [2–4]. A major structural characteristic of these complexes is the strength of their metal-ligand coordinative connectivity. With many transition metal ions in low oxidation states, a *bis*-complex is formed with a pseudo-octahedral coordination at the metal center. The inherent stability of this type of complex can be explained by the strong metal-ligand (d-π^*) back-donation as well as dynamic chelate effect.

Depending on the specific metal ion, complexes can show different stabilities, as expressed in the stability constants (K values), where K_1 represents the mono-complex (1 : 1) and K_2 the *bis*-complex (2 : 1) [5]. It can be concluded from the K values that Fe(II) complexes are more stable then the corresponding Mn(II) and Cd(II) complexes (Table 3.1).

---

[*] Parts of this chapter are reproduced from *Chem. Soc. Rev.* **2004**, *33*, 373-399 by permission of The Royal Society of Chemistry.

Modern Terpyridine Chemistry. U. S. Schubert, H. Hofmeier, G. R. Newkome
Copyright © 2006 WILEY-VCH Verlag GmbH & Co. KGaA, Weinheim
ISBN: 3-527-31475-X

Table 3.1 Stability constants of [Ru(tpy)$_2$]$^{2+}$ complexes measured in water [5].

Metal ion	log K_1	log K_2	log β_2
Mn^{2+}	4.4		
Fe^{2+}	7.1	13.8	20.9
Co^{2+}	8.4	9.9	18.3
Ni^{2+}	10.7	11.1	21.8
Zn^{2+}	6.0 6.7 [6]	5.2 [6]	
Cd^{2+}	5.1		

Table 3.2 Thermal stability (temperature of 5% weight loss in °C) of terpyridine ligands and metal complexes [7].

Terpyridine ligands		Bisterpyridine complexes			
H	5,5″-dimethyl	Mn^{2+}	Hg^{2+}	Co^{2+}	Zn^{2+}
225	245	315	336	371	390

Terpyridine ligands, as well as their corresponding metal complexes, were also investigated regarding their thermal stability utilizing thermogravimetric analysis (TGA). Compared to the free ligand, the metal complexes revealed a significant increase in thermal stability, as concluded from the temperature of onset of a 5% weight loss (Table 3.2) [7].

The common geometry of these *bisterpyridine* complexes is a distorted octahedral geometry, the usual coordination for transition metal ions being hexacoordinate. The distorted octahedral coordination geometry has been determined in detail by a single-crystal X-ray structure analysis of [Ru(tpy)$_2$(PF$_6$)$_2$] · 2 MeCN [8]. Figure 3.1 shows the crystal structure of a typical [Ru(tpy)$_2$]$^{2+}$ complex [9].

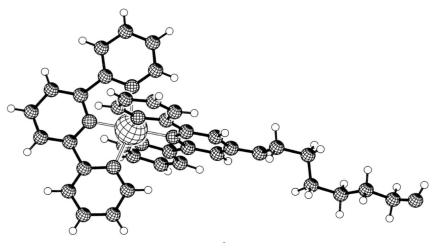

Figure 3.1 Crystal structure of a typical [Ru(tpy)$_2$]$^{2+}$ complex [9].

In order to obtain $[Ru(tpy)_2]^{2+}$ complexes, the ligand is treated with the specific metal ion [e.g., Zn(II), Co(II), Cu(II), Ni(II), Fe(II)] in a 2 : 1 (ligand : metal) ratio (Scheme 3.1a). The complexes are purified in a simple two-step procedure: (1) exchange of the counterions and (2) subsequent recrystallization. Addition of metal salts to a mixture of two different terpyridines leads to a statistical mixture of homo- and hetero-complexes. In order to prepare exclusively the asymmetric complexes, a directed strategy, where the two ligands are introduced in a two-step process (Scheme 3.1b) has to be used. Suitable metals for this strategy are generally Ru(III) and Os(III) (as well as Co(III) and Rh(III) [10]) in which the metal(III) 1 : 1-complex is isolated and subsequently reduced in situ (details are given below in Section 3.2.1.1). Strongly electron-withdrawing substituents, e.g., 4-cyano-tpy, have been noted to present problems in the construction of the desired $[Ru(tpy)_2]^{2+}$ complexes; recently however, the heteroleptic and homoleptic Ru(II) complexes have been reported [11].

The UV-vis spectra of terpyridine metal complexes show a bathochromic shift of the ligand-centered (LC) absorption bands for all metal complexes (Figure 3.2). In the case of the Fe(II) and Ru(II) complexes, a characteristic metal-ligand-charge-transfer (MLCT) band can be observed. This absorption lies in the visible region and is responsible for the intense color [purple for Fe(II) and red for Ru(II)]. For Fe(II) complexes, a distinct metal-centered band can be detected; whereas for Ru(II), only a shoulder was detected.

MALDI-TOF mass spectrometry is a well-suited tool for the analysis of *bis*terpyridine metal complexes [12]. Compared to other mass spectrometry

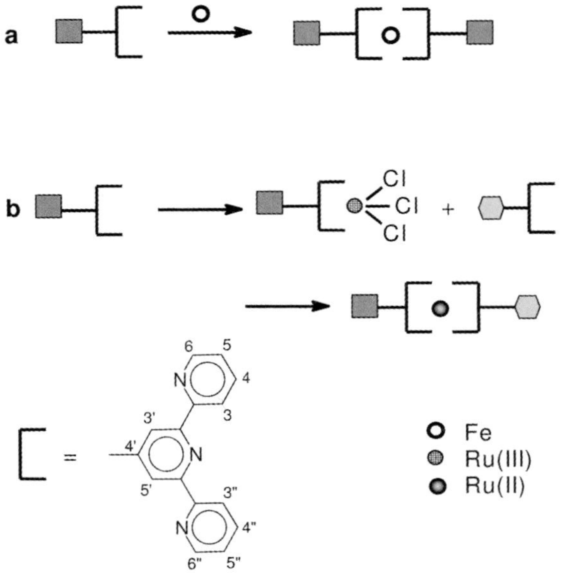

Scheme 3.1 Schematic representation of the formation of symmetrical (a) or unsymmetrical (b) *bis*terpyridine complexes.

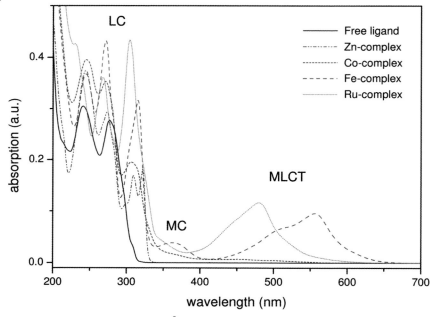

Figure 3.2 UV-vis spectra of $[M(tpy)_2]^{2+}$ complexes [Zn(II), Co(II), Fe(II), and Ru(II)] compared to the free ligand (in MeCN).

techniques, MALDI-TOF MS is very soft, thus allowing the ionization and detection of terpyridine complexes. The complex cation can be detected without ligand fragmentation. Usually complexes without counterions as well as ion pairs with one PF_6^- ion and sometimes with both counterions can be detected, in which the cation (without counterions) shows the highest intensity. In addition, matrix adducts could be observed for *bis*terpyridine metal complexes but are rare for the Ru(II) complexes. Furthermore, adducts of the cations from the dopant salt (Na^+ or K^+) can also be detected; however, at higher laser intensity, fragmentation of the complex cation occurred. The degree-of-fragmentation is dependent on the applied laser energy and thus allows an estimation of the complex's binding strength [12]. All species are singly, positively charged. This phenomenon can be explained by a plume of charged particles emitted in the desorption process that also contains electrons. Subsequently, a partial neutralization occurs from which the singly charged species survives the longest; therefore, only these ions were detected (the surviving more highly charged species have such a low intensity that they generally are not observable) [13, 14]. A typical MALDI-TOF MS is shown in Figure 3.3, also the comparison of the observed isotopic distribution pattern (left insert) with that of the simulation (right insert) revealed an excellent correlation.

The ^{1}H NMR spectra for the ligand and related *bis*-complex show a characteristic shift for the 6,6″-proton signals. The cause of this behavior is inherent in the

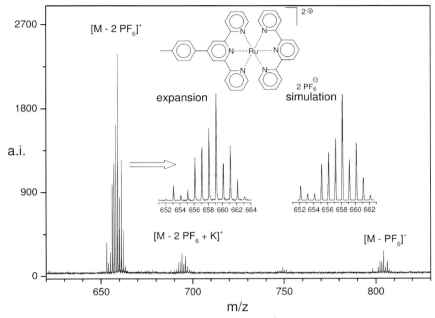

Figure 3.3 MALDI-TOF MS of an asymmetric [Ru(tpy)(tpy')]²⁺ complex including a simulation (right insert) of the isotopic pattern (see [9]).

configuration within the complex, in that the ligands are perpendicular to each other; thus, the 6,6"-protons in the *bis*-complex are located above the ring plane of the aromatic ring of the adjacent ligand, causing the observed upfield shift of these unique protons. Also the signals for the other protons reveal less dramatic shifts when compared to the free ligand. As well as the influence of the metal-ligand bond, the chemical environment is also different, in that the free ligand's nitrogen atoms possess an anti orientation in the crystal structures but in the complexes orientation must be syn for N-coordination (Figure 3.4).

Cobalt complexes show an interesting feature in that, despite their inherent paramagnetism, the ¹H NMR spectrum appears typically well resolved and the chemical shifts are influenced by the paramagnetic character of the Co(II) core. Due to hyperfine interaction of the unpaired Co(II) electrons with the ligand protons, a shift to low field was observed, and the signals were still sufficiently sharp for proton integration [15]. In that these Co(II) complexes can be either high-spin or low-spin, Constable et al. reported that the spin state of the Co(II) center can be distinguished by the chemical shifts of the NMR signals [16]. The lowest resonance in low-spin complexes can be found at ca. 100 ppm; however, it can be detected at ca. 250 ppm for high-spin complexes. Note that there are exceptions, such as bromoterpyridine, where the ligand field is weakened by the substituent [17, 18]. Figure 3.5 shows a typical ¹H NMR spectrum of a low-spin cobalt complex [19].

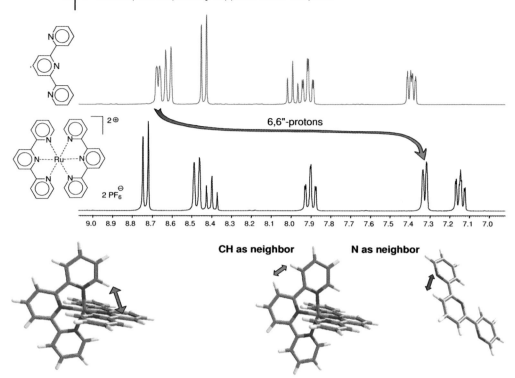

Figure 3.4 ^{1}H-NMR spectrum (a) of terpyridine and [Ru(tpy)$_2$(PF$_6$)$_2$] (in CD$_3$CN); (b) the critical steric interactions are depicted in the models.

Copper(II) ions allow the preparation of a special type of complex; thus, by using equimolar amounts of terpyridine and bipyridine, complexes containing both ligands can be obtained in one step [20, 21]. A two-step reaction also gave this type of complex [21], and the asymmetric complexes can be obtained in nearly quantitative yields. Helical trinuclear complexes could be synthesized by assembling a *tris*terpyridine ligand and a *tris*bipyridine ligand with Cu(II) ions [22]. The crystal data for the copper complexes revealed a pentacoordinated metal center of square-pyramidal geometry in which the sixth position is occupied by a loosely coordinated PF$_6^-$ counterion (Figure 3.6). A family of acetylide- and carbine-ruthenium complexes possessing both the terpyridine and bipyridine ligands has appeared; emission studies have shown some to be non-emissive while others produced an emission at diminished temperatures, which were assigned as d$_\pi$(Ru(II)) → π* (polypyridine) [3]MLCT in nature [23]. Other related Ru(II) complexes have recently appeared and were supported by X-ray analysis: [Ru(MeCN)(phenanthroline)(tpy)(PF$_6$)$_2$] [24], [Ru(tpy)(acac)<(4-hydroxy-3-methylbutynyl)phenyl-cyanamide>] [25], [Ru(typ)(2,6-*bis*(2-naphthyridyl)pyridine)(PF$_6$)$_2$] [26], [Ru(tpy)<1-[6-(2,2'-bipyridinyl)]-1-(2-pyridinylethanol>(ClO$_4$)$_2$] [27], and [Ru(tpy)(3-amino-6-(3,5-dimethylpyrazol-1-yl)-1,2,4,5-tetrazine)(Cl)] [28].

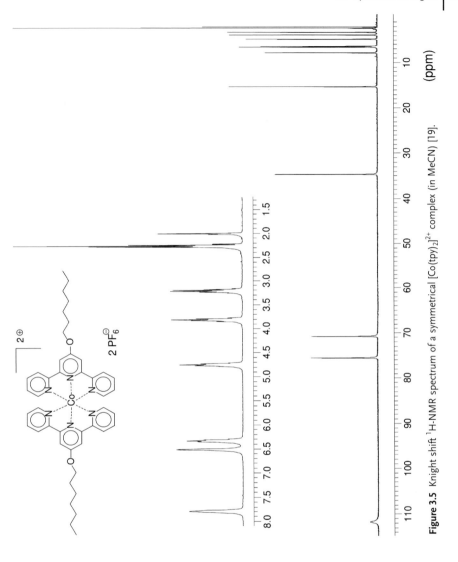

Figure 3.5 Knight shift ^{1}H-NMR spectrum of a symmetrical [Co(tpy)$_2$]$^{2+}$ complex (in MeCN) [19].

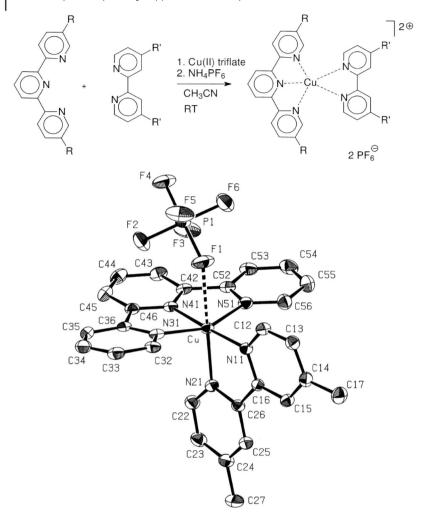

Figure 3.6 Schematic representation of the synthesis of bipyridine-terpyridine Cu(II) complexes; X-ray structure of such a typical complex. (Reprinted with permission from [29]).

3.2.1.1 Ruthenium Complexes

Complexation of terpyridine ligands with an Ru(II) center can be conducted in a simple two-step sequence (Scheme 3.2). The Ru(III) intermediate is generally isolated but not characterized, then simply reacted in the next step with a second equivalent of the same or different terpyridine under reductive conditions to afford the desired symmetric or asymmetric Ru(II) complex, respectively [3].

In the first step, $RuCl_3 \cdot nH_2O$ is added to either a methanolic or an ethanolic solution of the first terpyridine ligand. The resulting mono complex is generally insoluble and, in most cases, can be simply filtered; since it is paramagnetic,

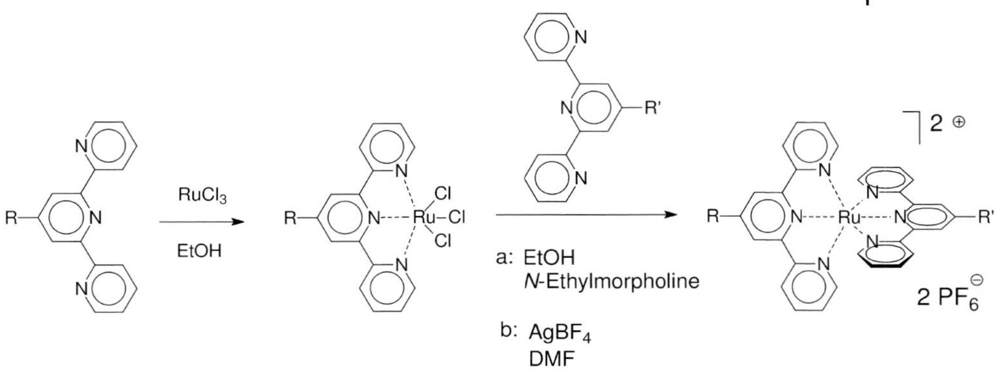

Scheme 3.2 Synthetic strategy toward asymmetric [Ru(tpy)(tpy')]$^{2+}$ complexes [3, 30].

typical NMR characterization cannot be conducted. This intermediate is subsequently suspended with the second ligand in methanol containing *N*-ethyl-
morpholine [31] and refluxed for 1–4 h. The solvent can also act as a reducing
agent for the Ru(III) → Ru(II) conversion. Alternatively, an equimolar amount of
AgBF$_4$ can be added to the initial Ru(III) complex in DMF or acetone to remove
chloride ions [30]. The vacant coordination sites are then occupied by the weakly
binding solvent molecules, affording an activated species that is soluble, unlike
the original [Ru(tpy)(Cl)$_3$] complex. This intermediate is reacted without isolation
with the second terpyridine (after filtration of the AgCl precipitate) to lead to the
desired [Ru(tpy)$_2$]$^{2+}$ complex; yields are typically in the range 50–90%. Although
this is a directed method, the formation of a statistical mixture of hetero- and
homo-complexes has been reported [32], but this is atypical behavior and was not
explained by the authors.

One of the mildest procedures for the preparation of [Ru(tpy)$_2$]$^{2+}$ complexes
[33] utilizes [Ru(tpy)(DMSO)(Cl)$_2$] [34]. Some functionalized terpyridines possessing structurally sensitive groups, e.g., ethynyl, can only be introduced by means
of this mild reagent. The structure of the precursor was confirmed by X-ray analysis.
Another (somewhat unusual) method was presented by Greene et al., who
demonstrated that microwave-assisted heating gave the desired product in yields
up to 94% within one minute [35].

Symmetric terpyridine complexes can be obtained by the RuCl$_3$-*N*-ethylmorpho-
line method, either in a two-step reaction by applying the same ligand twice or in
a one-pot reaction. The availability [36] of [Ru(DMSO)$_4$(Cl)$_2$] should also be
considered. Another method described by Rehahn starts from RuCl$_3$, which is
dechlorinated with AgBF$_4$ in acetone [37]. The resulting [Ru(Me$_2$CO)$_6$]$^{3+}$ complex
possessing the weakly bound solvent ligands is subsequently subjected to the
terpyridine ligands under reductive conditions.

It should be noted that the [Ru(tpy)$_2$]$^{2+}$ center can be reduced by electro-
crystallization, resulting in neutral complexes. The crystal structures of both
[Ru(tpy)$_2$]$^{2+}$ and [Ru(tpy)$_2$]^0 complexes have been accomplished showing indeed
that the latter neutral ruthenium complex does not possess external counter-ions

[38]. The observed species actually exists as an Ru(II) complex, but is coordinated to two terpyridine radical anions.

Photophysical Properties

Absorption, as well as emission spectra, revealed that a metal-ligand charge transfer takes place in these $[Ru(tpy)_2]^{2+}$ complexes. As opposed to bipyridine complexes, where a phosphorescence phenomenon can be observed over a wide temperature range, no emission is detected at ambient temperature in the case of terpyridines because of a non-radiative transition of the excited triplet metal-to-ligand charge-transfer (^{3}MLCT) state by a triplet metal-centered (^{3}MC) state to the ground state (see, e.g., [2]). At low temperatures, however, this path becomes less efficient, and luminescence can thus be observed [39–41]. For a sufficient observation of the emission properties, the material can be "dissolved" in a rigid glass at 77 K [42].

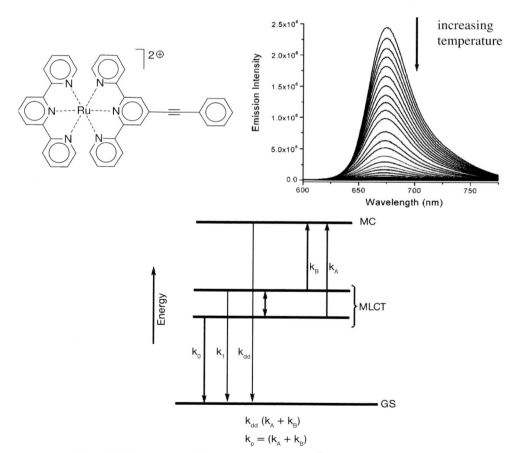

Figure 3.7 Top: structure of the $[Ru(tpy)(4'-PhC{\equiv}C-tpy]^{2+}$ complex and temperature dependent emission spectra (room temperature to 80 K), bottom: its energy level diagram. (Reprinted with permission from [43], © 2004 American Chemical Society).

Photophysical properties can, however, be fine tuned by the introduction of a donor, an acceptor, or both, leading to room-temperature luminescent $[Ru(tpy)_2]^{+2}$ complexes. Another method to improve the luminescence lifetime and quantum efficiency involves the attachment of aromatic groups. One of the first examples is represented by 4'-(p-tolyl)terpyridine; an extension of this approach by connecting aromatic rings to the terpyridine moiety leads to 4,4',4''-triphenylterpyridine, which revealed further improved optical properties. For more detailed information on these kinds of systems, the reader is referred to a detailed review by Balzani, Sauvage et al. [2].

The influence of delocalization was studied in detail on a mono(phenylethynyl)-*bis*terpyridine Ru(II) complex [43]. The luminescence of this room-temperature luminescent complex increases significantly at lower temperatures (Figure 3.7). A comparison with the non-luminescent unfunctionalized $[Ru(tpy)_2]^{2+}$ complex led to the conclusion that delocalization reduces the electron-vibrational coupling to the metal-centered state (responsible for radiationless deactivation).

A very different approach to the preparation of ambient-temperature luminescent ruthenium complexes is the entrapment of these materials in zeolite cages [44]. The zeolite-induced destabilization of the ligand-field (LF) state results in a strongly enhanced room-temperature emission of the $[Ru(tpy)_2]^{2+}$ complex. A feature article by Armaroli highlights these photophysical processes in various metallo-supramolecular compounds, including energy and electron transfer processes [45].

3.2.1.2 Other Luminescent Metal Complexes

Iridium(III) ions were found to show an enhanced luminescence behavior when compared to their Ru(II) analogues [46, 47]. Lifetimes above 1 μs could be obtained, which were assigned to be predominantly ligand-centered π-π* phosphorescence emissions [the metal-centered (MC) and metal-ligand-charge-transfer (MLCT) levels are at higher energies] [48]. Furthermore, asymmetric complexes could be synthesized in a similar fashion to that of the Ru(II) complexes: $IrCl_3 \cdot$ hydrate was treated with a terpyridine ligand to afford the Ir(III)-*mono*-complex. This intermediate was subjected to the second ligand without changing the oxidation state of the metal center, but under harsher reaction conditions (diethylene glycol at 200 °C). Among the examples are complexes containing both a tolyl and a pyridinyl group; the latter was subsequently converted into a N-methylpyridinium moiety. The luminescence of these complexes was quenched in the presence of chloride ions, making such systems interesting sensors [49]. A recent contribution from Williams et al. describes *bis*(4'-biphenylterpyridine) iridium complexes, which exhibit intense long-lived emission due to the added aromatic ring conjugation. This explanation was confirmed by a comparison to analogous complexes of 4'-mesitylterpyridine, where conjugation is completely hindered because of steric repulsion [50]. Recently, a series of N-phenylbenzamide-substituted $[Ir(tpy)_2]^{3+}$ complexes have appeared [51] and have been shown to possess ^3CT emission properties causing a low-energy shift of the main absorption bands when compared to $[Ir(tpy)_2]^{3+}$.

Iridium complexes bearing porphyrins and cyclodextrins will be described in a later section. For a recent review on such iridium-systems, see also [52].

Osmium complexes of terpyridines are also luminescent at room temperature. The radiationless deactivation is less efficient because of a higher energy gap between the triplet MLCT state and the triplet MC state: the ^{3}MLCT level of Os(II) complexes lies at a lower energy than Ru(II), since (a) Os(II) is easier to oxidize and (b) the ^{3}MC state lies at higher energy, the ligand field of Os(II) being stronger than that of Ru(II). Osmium complexes, especially in dyads with ruthenium complexes, will be considered later. (Asymmetric complexes can be synthesized in a manner analogous to their Ru(II) counterparts but not without statistical ligand scrambling – there are, however, gentle methods available to make the unsymmetrical Os(II) complexes [53].) Another metal ion, leading to luminescent species, is that of platinum [54], which forms [Pt(tpy)(Cl)]$^+$ type complexes [55]. Whereas the parent complex is non-luminescent at room temperature, complexes of 4'-substituted terpyridines or different co-ligands are emitters. The DNA-binding properties [56–58] of these Pt-complexes make them interesting sensors in biochemistry [59, 60]. Chromium(III) complexes [Cr(ttpy)$_2$(ClO$_4$)$_3$] and [Cr(Brphtpy)$_2$(ClO$_4$)$_3$], where ttpy = 4'-(p-tolyl)terpyridine and Brphtyp = 4'-(p-bromophenyl)terpyridine [61], have been prepared, characterized, and shown to be moderate binders of calf thymus DNA; the influence of DNA on the emission of these complexes has been described [62].

3.3
Mononuclear *Bisterpyridine* Ruthenium Complexes

Considerable work has already been done in the field of terpyridine complexation, mostly on mononuclear terpyridine complexes, especially regarding the fine-tuning of their optical properties or their use as precursors for the construction of supramolecular architectures.

Carboxy groups (Figure 3.8) play an important role as substituents on terpyridine ligands because of their potential use as surface anchoring groups or potential internal counterions. For example when [Ru(tpy)$_2$]$^{2+}$ complexes are attached to TiO$_2$, which has been widely used in solar cell applications [63], novel solar cells are possible.

The electropolymerizable thienyl group has also been introduced (Figure 3.8a) into a terpyridine complex [64, 67]; the {[RuLL'][PF$_6$]$_2$ (L = 4'-(3-thienyl)terpyridine; L' = 4'-carboxyterpyridine)} complex was prepared by two different synthetic pathways. The first involved a protected carboxylic acid moiety during complexation followed by hydrolysis, since direct complexation with the free carboxylic acid resulted in a very low yield. The second approach was based on the oxidation of a furan ring to generate the carboxyl moiety and was shown to be a superior approach; thus, 4'-(2-furyl)terpyridine was reacted with 4'-(3-thienyl)terpyridine ruthenium trichloride [68, 69], followed by permanganate oxidation under basic conditions, affording the desired carboxy product.

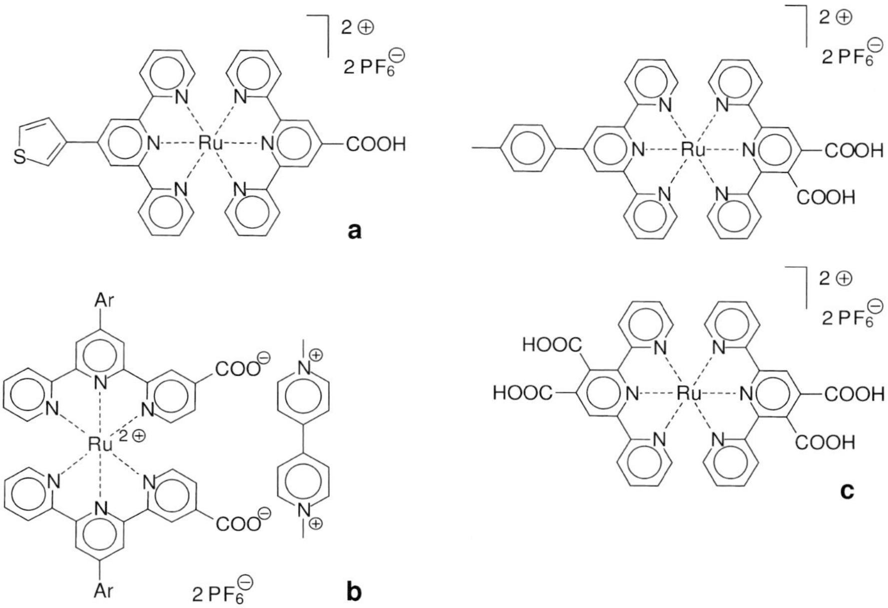

Figure 3.8 Carboxy-functionalized [Ru(tpy)$_2$]$^{2+}$ complexes [64–66].

Carboxy groups have also been introduced into the 4-position of terpyridines (Figure 3.8b) [65]. Because of their spatial orientation, they can easily coordinate to methyl viologen [4,4'-*bis*(methylpyridinium)] by ionic interactions, giving rise to a photoinduced electron transfer.

In order to increase the electron injection efficiency and to reduce the chance of complex desorption from a TiO$_2$ surface, asymmetric [Ru(tpy)(tpy')]$^{2+}$ complexes possessing vicinal carboxylic acids were developed (Figure 3.8c) [66]. Because complexation of the free diacid resulted in the loss of one carboxylic group (decarboxylation), initial esterification was necessary. After saponification, the resulting complexes demonstrated room-temperature luminescence and efficient sensitization of nanocrystalline TiO$_2$ films, with conversion yields (incident photon to current efficiency, IPCE) of up to 70%.

A new versatile preparation of boron-functionalized terpyridines and their corresponding metal complexes has been reported [70]. In the following example, 4'-(4-bromophenyl)terpyridine [61] was functionalized with boronate groups by a Pd-catalyzed Miyaura cross-coupling, resulting in the first example of a boronate-functionalized terpyridine (Scheme 3.3). Subsequent hydrolysis of this intermediate led to the free boronic acid. The ligand, bearing the boronic ester, was successfully complexed with Ru(II); while the corresponding free acid disintegrated under these reaction conditions because of the destabilized boron bond. Interestingly, the boronation could also be performed on a preformed Ru(II) complex. The resulting boronate complexes were useful starting materials for further reactions involving the boronate group (e.g. Suzuki-Miyaura cross-couplings).

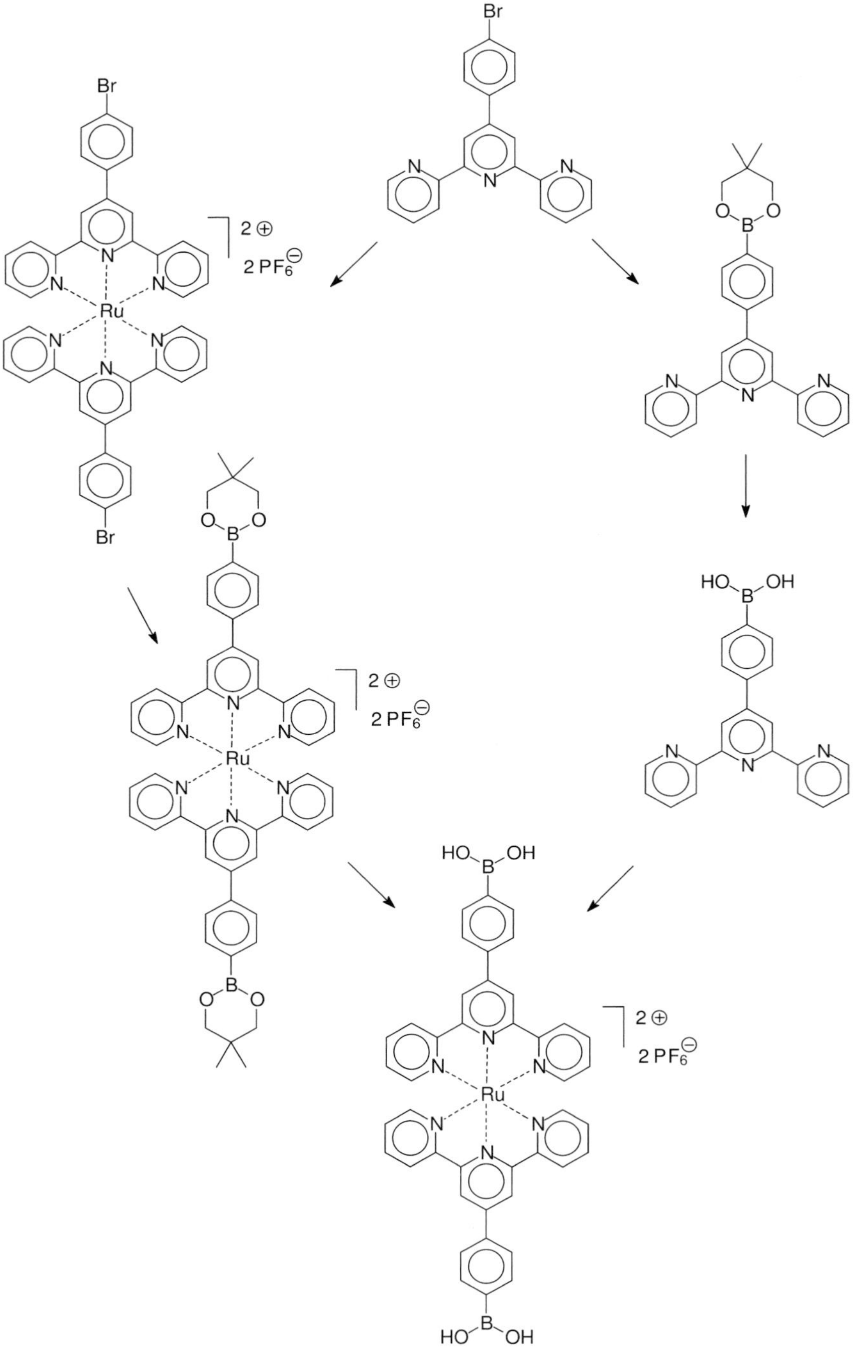

Scheme 3.3 Synthesis of boronate [Ru(tpy)$_2$]$^{2+}$ complexes [70].

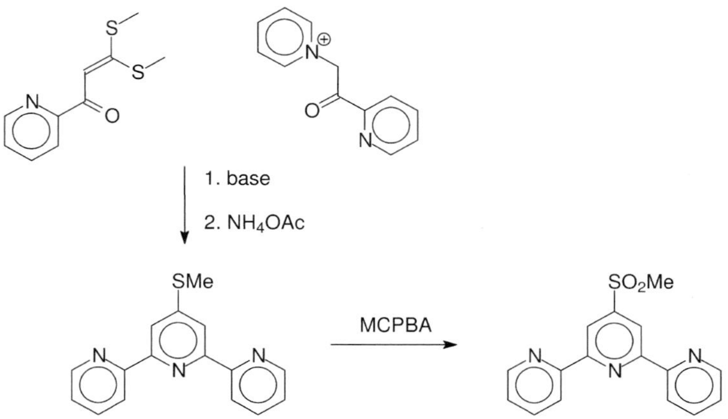

Scheme 3.4 Synthesis of a terpyridine sulfonate [71].

Electron-withdrawing groups in the terpyridine 4′-position have been shown to afford room-temperature luminescent Ru(II) complexes [31]. Since 4′-chloro-terpyridine revealed a weak luminescence, and in order to obtain a more efficient complex, a 4-methylsulfone group was introduced via a 4′-(methylthio)terpyridine (Scheme 3.4) [71], which was prepared according to the Potts procedure [72] and then oxidized with a peracid to afford the desired sulfone.

Another acceptor moiety that has recently been reported [73] is that of a triphenylpyridinium group; however, because of the steric demands of the aromatic rings, substitution-extended conjugation is not possible, since molecular rotation is inhibited (Figure 3.9a). In spite of the lack of conjugation between the terpyridine

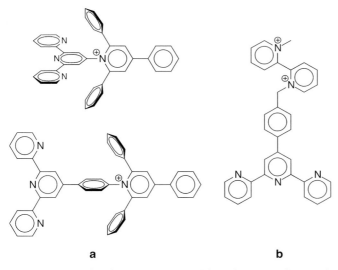

Figure 3.9 (a) Triphenylpyridinium- [74] and (b) viologen- [75] functionalized terpyridine ligands.

and triphenylpyridinium moiety, an enhanced room-temperature luminescence was still observed [74]. This behavior was ascribed to a through-bond electronic substituent effect originating from the directly connected electron-withdrawing group. In the elongated case, where the unit is separated by a *para*-phenyl spacer, the formation of photoinduced electron-transfer (PET) processes with the formation of charge-separated (CS) states is possible.

Another novel example, in which the pyridinium moiety was connected to a terpyridine, is shown in Figure 3.9b. In this case, 1-methyl-1′-[4-(2,2′:6′,2″-terpyridin-4-yl)benzyl]-4,4′-bipyridinium was used to prepare the corresponding 2 : 1-osmium complex, in which the characteristic osmium luminescence was efficiently quenched by the viologen moiety. The quenching probably involves an intramolecular electron transfer [75].

Besides the introduction of electron-withdrawing groups [11], the extension of the ligand's conjugated π-system by one or more (hetero)aromatic rings was also a promising approach (Figure 3.10).

One strategy toward such systems is by the attachment of (hetero)aromatic rings. In the case shown in Figure 3.10a, a pyrimidine moiety was introduced into the 4′-position of a terpyridine [76]; further substitution of that initial pyrimidine at its 4-position with a second pyrimidine, phenyl, or pentafluorophenyl groups extended the conjugated π-system. This structural elongation led to an increased gap between the metal-centered and MLCT states, resulting in room-temperature (lifetime up to 200 ns) luminescence. Other examples of luminescent ruthenium complexes include complexes involving (2-aryl) [79]- or 2-(4′-pyridyl)-4,6-*bis*-(2′-pyridyl)-1,3,5-triazine. These *bis*(2′-pyrimidyl)-1,3,5-triazines can be considered as "heteroterpyridines" (nitrogen atoms in the 3,3′,5′,3″-positions); many are capable of possessing room-temperature luminescence (Figure 3.10b) [77]. These complexes show luminescence due to a coplanar arrangement of the aromatic rings, when *N*-heteroatoms are present in the ortho position (Figure 3.11).

In view of the repulsion of the *ortho*-hydrogen atoms (Figure 3.11), the phenyl rings directly attached to a terpyridine are not coplanar. Similar [Ru(tpy)$_2$]$^{2+}$ complexes derived from 4′-pyridyl- and 4′-pyrimidyl-functionalized terpyridines, and *N*-heteroterpyridines have been prepared and characterized (Figure 3.10c) [78]; however, despite the similarity to the previously described systems, these authors reported that these complexes are non-luminescent.

An interesting feature of 2,4,6-*tris*(2-pyrimidyl)-1,3,5-triazine is that it is actually composed of three terpyridine subunits; the reported complex thus contains two vacant complexing units. This architectural feature makes this system interesting for the construction of extended supramolecular assemblies as will be described later.

Besides simple (hetero)aromatic attachments, fused-aromatic hydrocarbons have been introduced into terpyridines, in which the corresponding ruthenium complexes showed room-temperature luminescence. Among the notable examples is the 4′-(9-anthryl)terpyridine (An-tpy), which has been complexed to form [Ru(An-tpy)$_2$]$^{2+}$ and [Fe(An-tpy)$_2$]$^{2+}$ complexes [80]. In a later report, this ruthenium complex was reported as having no luminescence, since excitation of the complex

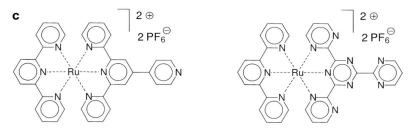

a

b

R = —Cl

—C₆F₅
—CN

c

Figure 3.10 Phenyl- and heteroaryl-functionalized complexes of heteroterpyridines [76–78].

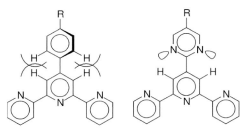

Figure 3.11 Comparison of the conformations of phenyl- and pyrimidine-substituted terpyridines.

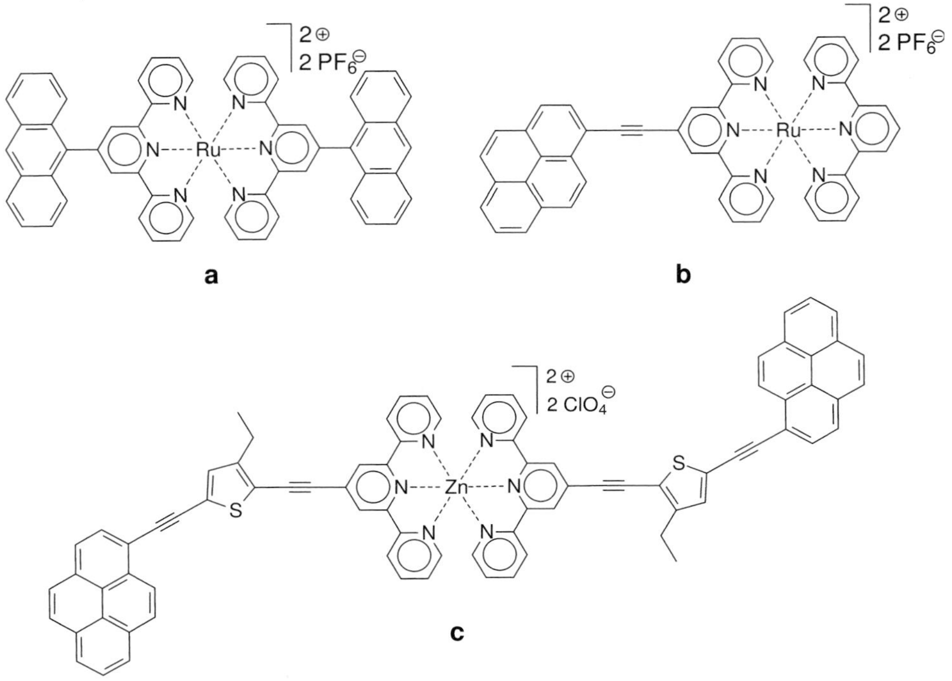

Figure 3.12 [Ru(tpy)$_2$]$^{2+}$ complexes, containing
(a) 9-anthryl- [82],
(b) pyrene-1-ylethynyl- [83], or
(c) 4-ethyl-5-pyren-1-ylethynylthiophen-2-ylethynyl- [84] extended aromatic functions.

can cause the formation of the triplet excited state of anthracene, which is lower in energy than the ^{3}MLCT of [Ru(tpy)$_2$]$^{2+}$ and is non-emissive [81]. The species [Os(An-tpy)$_2$]$^{2+}$ and [Zn(An-tpy)$_2$]$^{2+}$ were also made [81]; the Os complex was luminescent. Other examples include the introduction of 1-pyrenyl moieties (Figure 3.12) [82]. Recently, ethynylpyrene units have also been introduced in the 5,5″-positions of a terpyridine [83]. Moreover, pyrene moieties have been linked by means of a (*bis*-ethynyl)thiophene bridge to the terpyridine [84]. Recently, a series of [Ru(tpy)$_2$]$^{2+}$ and [Os(tpy)$_2$]$^{2+}$ complexes possessing bithienyl, quaterthienyl, or hexathienyl moieties have been prepared and shown to form rod-like polymers [85]; the dc conductivity of the metal-based redox couple is 2 orders of magnitude greater than that of a comparable non-conjugated system.

The initial ligand to the complex shown in Figure 3.12c is luminescent at room temperature due to an efficient population of an intramolecular charge transfer excited state; however, addition of Zn(II) ions resulted in quenching of the emission, making such systems interesting metal ion "sensors". Furthermore, alkyne cross-coupling reactions have been used to obtain fluorene-functionalized terpyridine *bis*-complexes [86], in which fluorene units were coupled to terpyridines

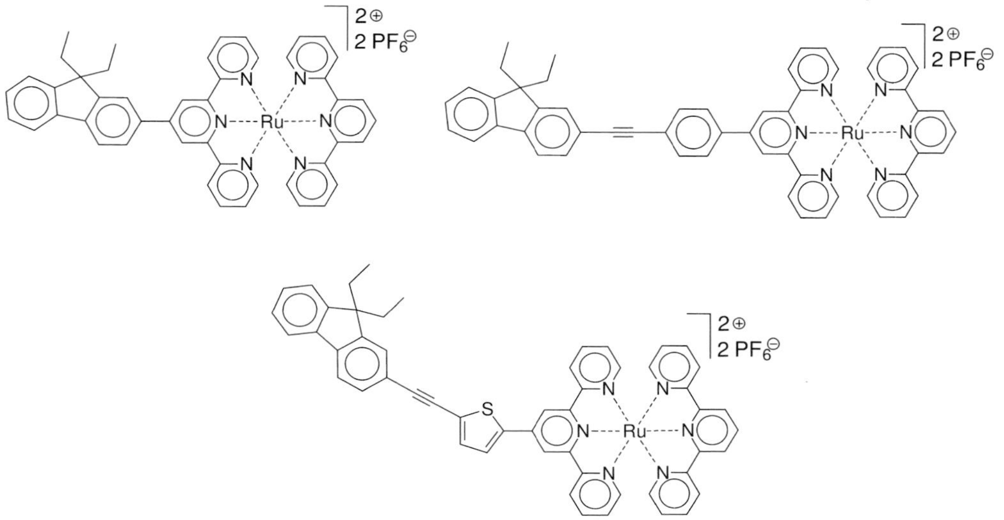

Figure 3.13 Fluorene-functionalized $[Ru(tpy)_2]^{2+}$ complexes [86].

directly or via phenylene-ethynylyne or thiophenylethynylyne connectors applying Kröhnke synthesis and Sonogashira coupling reactions. The Ru(II) complexes, formed from these ligands (Figure 3.13) are non-luminescent despite the attached conjugated groups, whereas the corresponding Zn(II) complexes are highly luminous.

A rather new concept for elongating the luminescence of $[Ru(tpy)_2]^{2+}$ complexes was described as the "multichromophore approach". Organic chromophores with suitable triplet states (long-lived because their deactivation is forbidden), which are close to the ^{3}MLCT state, could act as "storage" for the excitation energy, resulting in prolonged lifetimes of these states [87]. Pyrimidine-substituted terpyridines with extended π^* orbitals were used as ligands to which anthracene moieties were connected as organic chromophores (Figure 3.14). This effect could also be shown when the "storage" unit was not directly connected to the ligand involved with the MLCT excited state [88]; however, when anthracene was connected to a $[Ru(tpy)_2]^{2+}$ complex (without pyrimidine unit), the triplet energy was quenched by the anthracene moiety.

As well as the photophysical properties originating from this delocalization phenomena, the intermolecular interactions of such compounds were also found to be of interest. Extended aromatic systems are known to give rise to $\pi–\pi$ interactions, which could be relatively strong for large molecules. Such π-stacking could be exploited for the 3-dimensional arrangement of terpyridine complexes (Figure 3.15).

The *Alcock* group reported the synthesis of 4′-biphenyl-functionalized ter-pyridines as well as the preparation of their corresponding *bis*-complexes using a variety of metal ions [89]. Furthermore, a ruthenium *bis*-complex with one biphenyl-

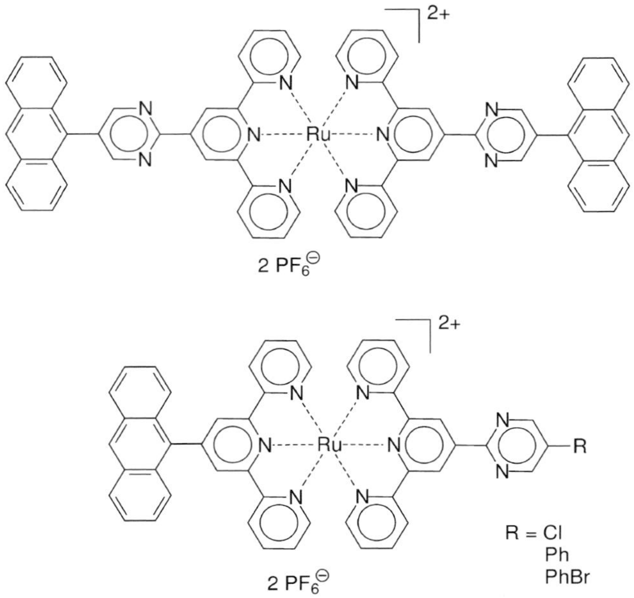

Figure 3.14 Anthracene- and pyrimidinyl-[Ru(tpy)$_2$]$^{2+}$ complexes [87, 88].

M = Co, Ru, Ni, Cu, Zn, Cd

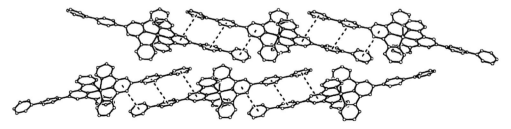

Figure 3.15 π-Stacking of 4′-(4-biphenyl)-substituted [Ru(tpy)$_2$]$^{2+}$ complexes [89]. (Reproduced by permission of The Royal Society of Chemistry).

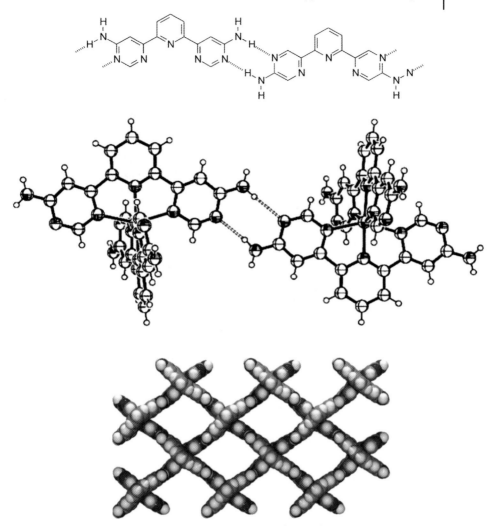

Figure 3.16 Schematic representation and crystal structure of terpyridines with H-bonding groups that assemble to a grid-like superstructure. (Reprinted with permission from [90]).

terpyridine and one unfunctionalized terpyridine was synthesized. π-Stacking interactions between the biphenyl and pyridinyl rings were observed in the crystal structure resulting in aggregation of the complexes in the solid state. Depending on the metal ion, different structures were found: for cobalt, ruthenium, nickel, and copper, biphenylene-biphenylene interactions led to linear rod-like arrays; whereas for complexes of zinc and cadmium, biphenylene-pyridyl interactions led to two-dimensional sheets. Moreover, ruthenium, zinc, and cadmium complexes showed room temperature luminescence both in the solid state and in solution.

Another non-covalent interaction, which has been used in the chemistry of terpyridine complexes, is *H*-bonding; in one particular example, two "hetero-terpyridines", containing amino-pyrimidine and amino-pyrazine moieties, were synthesized and subsequently shown to form double *H*-bonding arrays [90]. The solid-state structures did indeed reveal the formation of extended grid-like structures through molecular recognition (Figure 3.16).

While PF_6^- counterions of a cobalt *bis*-complex resulted in the complete saturation of the *H*-bonds, the corresponding BF_4^- counterions led to partially broken networks. With zinc triflate, only half of the bonds are formed, leading to a chain-like assembly. An explanation for these events is that each has a different packing within the crystal lattice relative to the size of respective counterions. These complexes could be considered as a prototype for the generation of novel organized arrays of terpyridine complexes through sequential self-assembly processes. Another example of a combination of *bis*-terpyridine ruthenium complexes and *H*-bonding interactions was based on the ureidopyrimidinone quadruple *H*-bonding unit, which afforded supramolecular dimers in apolar solvents [9].

Numerous differently functionalized terpyridine analogs, where the pyridine rings have been replaced by either quinoline [91] or phenanthroline moieties, have been used to form symmetric as well as asymmetric complexes [92]. Besides functional groups, connected by σ-bonds in various positions, fused-phenyl rings were also introduced (Figure 3.17). Among the species described, was a "terpyridine" consisting of a pyridylphenanthroline combination. Moreover, cyclo-metalated species have been prepared [92] in which one *N*-atom was replaced by carbon, resulting in a formal Ru-C bond. Some of these complexes also showed long MLCT excited-state lifetimes (70–106 ns) at room temperature.

With a view to developing applications in solar cells or artificial photosynthesis, photosensitizer-electron acceptor systems have been prepared by appending a naphthalenediimide moiety to the terpyridine (e.g., an aminoterpyridine with the naphthaleneanhydride) and subsequent conversion to the desired ruthenium complexes. Unsubstituted terpyridine as well as 4′-*p*-tolyl-terpyridine have been installed as the second ligand; the acceptor unit has been attached in a rigid fashion via either a phenylene group or an additional saturated carbon atom (Figure 3.18a) [93, 94]. However, unlike the corresponding bipyridine complexes, no electron transfer processes were detected in the case of the terpyridine complexes; this was thought to be a result of the short lifetime of the excited state of the $[Ru(tpy)_2]^{2+}$ complex.

Oligo(*p*-phenylenevinylene) (OPV) continues to be of special interest because of its outstanding optical properties. Furthermore, it is known to act as a donor in an efficient photoinduced electron transfer, which is of interest for solar cell applications. In combination with fullerene compounds acting as acceptors, the photochemically generated charges could be separated and drained with suitable compounds. In order to improve this system, two OPV units bearing terpyridine ligands (synthesized by a Pd-catalyzed cross-coupling reaction) have been converted to an $[Ru(tpy)_2]^{2+}$ complex (Figure 3.18b) [95]. Near steady-state photoinduced absorption experiments did indeed reveal a charge-separated

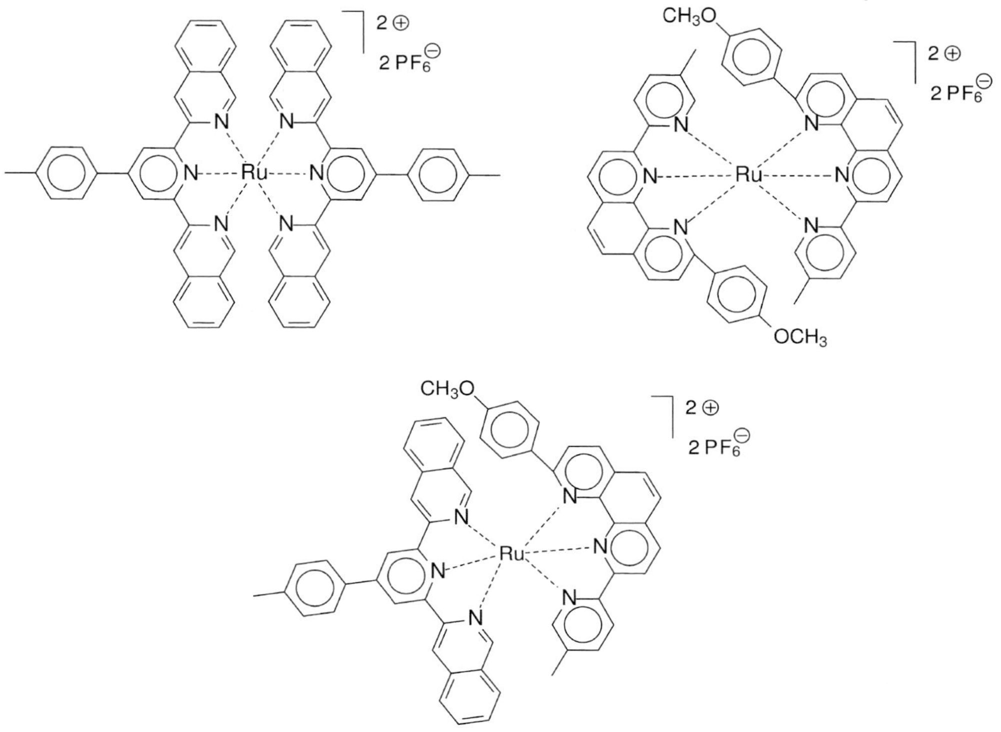

Figure 3.17 Ruthenium(II) complexes of "terpyridines" with fused aromatic rings [92].

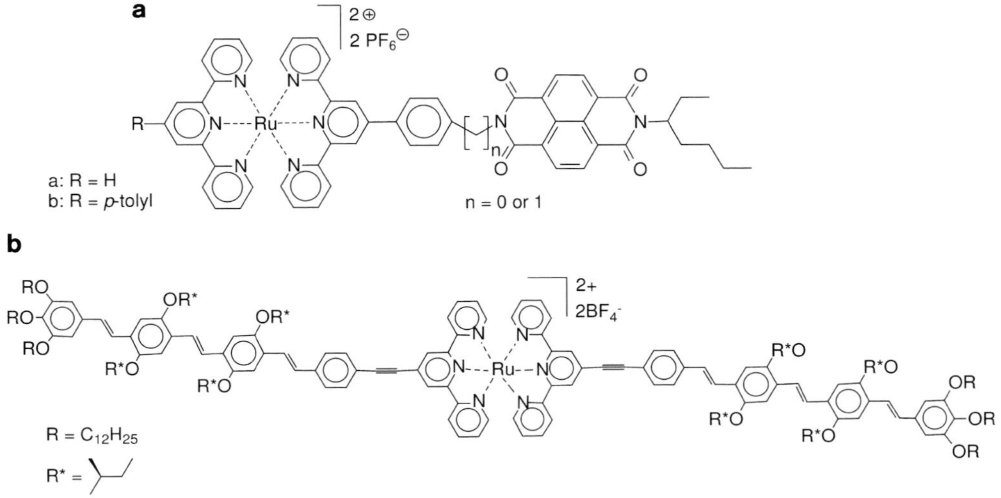

a: R = H
b: R = p-tolyl

n = 0 or 1

R = C$_{12}$H$_{25}$

R* =

Figure 3.18 Naphthalenediimide- [93] and oligo(p-phenylenevinylene)- [95] substituted [Ru(tpy)$_2$]$^{2+}$ complexes.

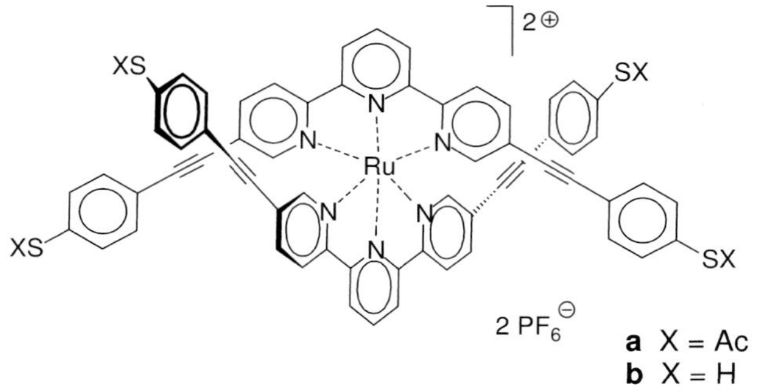

Figure 3.19 Conjugated thiol $[Ru(tpy)_2]^{2+}$ complex [96].

state, which was dependent on solvent polarity. In further experiments, the OPV donor has been coupled directly to a fullerene by terpyridine-ruthenium complexation.

Ruthenium complexes containing fully conjugated ligands terminated with thiol groups (5,5''-terpyridine substitution) have been obtained by cross-coupling of the corresponding alkynes. The resulting compounds were subsequently immobilized on gold surfaces (Figure 3.19). Self-assembled monolayers of single molecules have been imaged using scanning tunneling microscopy [96]. The delocalized phenyleneethynyl moiety could act as molecular wire, opening avenues into molecular electronics. The molecular inclusion of 1',1''-bis(terpyridinyl)biferrocene centers has also been recently suggested as a method to perturb the electronic properties of the spacer unit [97].

There are numerous examples where terpyridines were combined with tridentate terpyridine analogs (Figure 3.20). In the case of bis-triazole-pyridine, for example, different isomers are possible, because this ligand possesses different sets of nitrogen atoms that can coordinate the ruthenium ion [98]. Compared to $[Ru(tpy)_2]^{2+}$, the luminescence lifetime of the these systems showed a 300-fold increase, which was explained by a rise of the energy of the ^3MC level (in contrast to the previously described examples, where the ^3MLCT state was actually lowered). Furthermore, 5-phenyl-bis-triazole- and bis-tetrazole-pyridines have been reported; the triazole rings are N-bonded 6π electron ligands in these complexes and reprotonation resulted in quenching of the emission [99]. The accessible and related 2,6-bis(alkylimidazolium-3-yl)pyridine salts have been readily converted to the "pincer-type" (1 : 1) $[Ru(cnc)(CO)(Br)_2]$ and (2 : 2) $[Ru(cnc)_2(PF_6)_2]$ complexes, where cnc = 2,6-bis(butylimidazol-2-ylidene)pyridine, which was fully characterized [100]. The incorporation of the easily prepared 2,6-bis(1'-methylbenzimidazolyl)pyridine has been shown to offer a route to polymers with interesting optical properties [101]. Treatment of $[Ru(\eta_6\text{-cot})(dmfm)_2]$, where cot = 1,3,5-cyclooctatriene and dmfm = dimethyl fumarate, with terpyridine in refluxing acetone under nitrogen gave (36%) a mixture of stereoisomers of $[Ru(dmfm)_2(tpy)]$, which is the first example of an isolable mononuclear zerovalent ruthenium terpyridine complex. Similarly, treatment of $[Ru(\eta_6\text{-cot})(dmfm)_2]$ with 2,6-bis[(4S)-(-)-isopropyl-

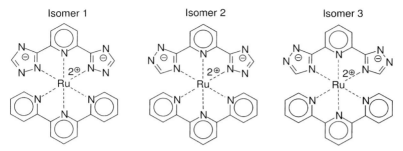

Figure 3.20 Different coordination isomers of the complex [Ru(tpy)(*bis*-triazole pyridine)] [98].

2-oxazolin-2-yl]pyridine (*i*-Pr-Pybox) afforded the corresponding [Ru(dmfm)$_2$(*i*-Pr-Pybox)] in good yields [102].

Attempts have been made to prepare liquid crystals containing metal complexes in order to obtain mesophases in which the characteristic properties of such complexes are incorporated. One approach to achieving this goal consisted of an amphiphilic terpyridine Ru(II) complex possessing an alkyl chain, e.g., C$_{19}$H$_{31}$ and C$_{31}$H$_{63}$, which could act as a surfactant [103]. For the complex containing the shorter chain, lyotropic mesomorphism was found in water, while similar behavior was detected for the longer-chain complex in ethylene glycol.

Recently, a new series of MRI contrast agents were synthesized [104] by the attachment of polyaminocarboxylate diethylene-*N,N,N″,N″*-tetraacetate (DTTA) [105] to the 4′-position of terpyridine, followed by assembly with Fe(II) and addition of Gd(III) to afford the desired [Fe(tpy-DTTA)$_2$Gd$_2$(H$_2$O)$_2$]; the Ru(II) complex was also constructed. Both of these complexes were shown [104] to possess a "remarkably" high relaxivity (r_1 = 15.7 and 15.6 mM^{-1} s^{-1} respectively).

3.4
Chiral Complexes

First attempts to introduce chirality into terpyridine metal architectures were undertaken by Abruña et al.; these structures were multinuclear helicates and are discussed in the next chapter. A simple mononuclear system was described [106] in which chirality was introduced by a 2,2-*bis*[2-(4(S)- or 4(R)-phenyl-1,3-oxazoli-nyl)]propane that was coordinated with unfunctionalized terpyridine to an oxo-ruthenium(IV) complex.

Von Zelewsky et al. cleverly utilized a chiral "dipineno"-(5,6:5″,6″)-fused 2,2′:6′,2″-terpyridine ligand as well as the analogous (4,5:4″,5″)-fused derivatives [32]. Here, the chirality was introduced by the fusion of a chiral pinene moiety onto the side terpyridine rings (Figure 3.21). The ligands were synthesized enantiomerically pure. Circular dichroism (CD) investigations showed that the resulting complexes are helically distorted in a chiral fashion. These and other pinene-functionalized terpyridines have been evaluated as potential stereoselective catalysts [107].

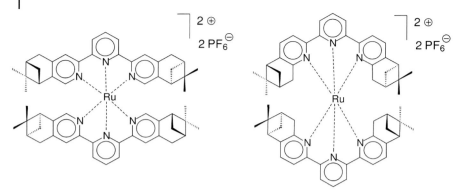

Figure 3.21 Chiral [Ru(tpy)$_2$]$^{2+}$ complexes with fused pinene units [32].

Furthermore, attempts were undertaken to introduce chirality into side groups which are connected only via one σ-bond rather than single fusion. Besides the examples of 6,6″-substituted terpyridines (Figure 3.22a,b) [107], Constable et al. [108] reported a 4′-bornylterpyridine (Figure 3.22c).

Chiral [Ru(tpy)$_2$]$^{2+}$ as well as [Co(tpy)$_2$]$^{2+}$ complexes were synthesized using ligands of this type; however, transfer of chiral information to the complex core was minimal. Schubert et al. prepared 4′-[3,4,5-*tris*(3,7-dimethyloctyloxy)phenyl-ethynyl]terpyridine via a Pd-mediated alkyne cross-coupling [109, 110]; the distance between the chiral groups and the terpyridine is larger than in the above described systems. The CD spectroscopy of its [Ru(tpy)$_2$]$^{2+}$ complex in an apolar solvent (i.e., dodecane) revealed notable chirality, suggesting that the transfer of chirality into the complex was by aggregation (Figure 3.23). In solvents where complete dissolution of the complex was possible, no CD effect was found.

Recently, Constable et al. constructed a *bis*-ligand, 2,7-di[terpyridinoxy-*bis*(ethyleneoxy)]naphthalene. When this was treated with Fe(II) salts, macrocyclization occurred, affording a new ditopic complex possessing both enantiomeric forms, as shown by single-crystal X-ray structure [111].

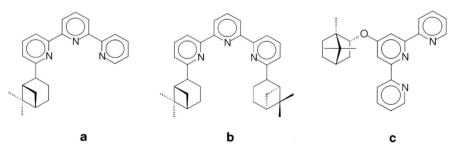

a　　　　　　　**b**　　　　　　　**c**

Figure 3.22 Connecting a chiral unit via a single bond [107, 108].

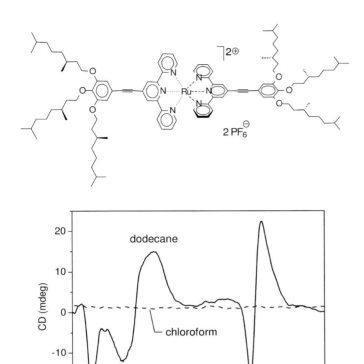

Figure 3.23 Structure of the [Ru(tpy)$_2$]$^{2+}$ complex bearing chiral side groups and CD spectra in chloroform and dodecane, respectively [109].

References

1 A. C. Benniston, *Chem. Soc. Rev.* **2004**, *33*, 573–578.

2 J. P. Sauvage, J. P. Collin, J. C. Chambron, S. Guillerez, C. Coudret, V. Balzani, F. Barigelletti, L. De Cola, L. Flamigni, *Chem. Rev.* **1994**, *94*, 993–1019.

3 E. C. Constable, M. W. C. Thompson, *New J. Chem.* **1992**, *16*, 855–876.

4 P. Tomasik, Z. Ratajewicz, *Pyridine Metal Complexes*, John Wiley & Sons, New York, **1985**.

5 R. H. Holyer, C. D. Hubbard, S. F. A. Kettle, R. G. Wilkins, *Inorg. Chem.* **1966**, *5*, 622–625.

6 P. R. Andres, PhD thesis, Eindhoven University of Technology (Eindhoven), **2004**.

7 U. S. Schubert, C. Eschbaumer, Q. An, T. Salditt, *J. Incl. Phen.* **1999**, *35*, 35–43.

8 K. Lashgari, M. Kritikos, R. Norrestam, T. Norrby, *Acta Cryst. C* **1999**, *C55*, 64–67.

9 H. Hofmeier, P. R. Andres, R. Hoogenboom, E. Herdtweck, U. S. Schubert, *Aust. J. Chem.* **2004**, *57*, 419–426.

10 J. Paul, S. Spey, H. Adams, J. A. Thomas, *Inorg. Chim. Acta* **2004**, *357*, 2827–2832.

11 J. Wang, Y.-Q. Fang, G. S. Hanan, F. Loiseau, S. Campagna, *Inorg. Chem.* **2005**, *44*, 5–7.

12 M. A. R. Meier, B. G. G. Lohmeijer, U. S. Schubert, *J. Mass Spectrom.* **2003**, *38*, 510–516.

13 M. Karas, M. Glückmann, J. Schäfer, *J. Mass Spectrom.* **2000**, *35*, 1–12.

14 U. S. Schubert, C. Eschbaumer, *Polym. Prepr.* **2000**, *41*, 676–677.

15 I. Bertini, C. Luchinat, *Coord. Chem. Rev.,* **1996**, *150*, 1–292.

16 H. S. Chow, E. C. Constable, C. E. Housecroft, K. J. Kulicke, Y. Tao, *Dalton Trans.* **2005**, 236–237.

17 E. C. Constable, C. P. Hart, C. E. Housecroft, *Appl. Organomet. Chem.* **2003**, *17*, 383–387.

18 E. C. Constable, C. E. Housecroft, T. Kulke, C. Lazzarini, E. R. Schofield, Y. Zimmermann, *J. Chem. Soc., Dalton Trans.* **2001**, 2864–2871.

19 H. Hofmeier, Ph.D. thesis, Eindhoven University of Technology (Eindhoven), **2004**.

20 G. Arena, R. P. Bonomo, S. Museumeci, R. Purello, E. Rizarelli, *J. Chem. Soc., Dalton Trans.* **1983**, 1279–1283.

21 C. M. Harris, T. N. Lockyer, *Aust. J. Chem.* **1970**, *23*, 673–682.

22 B. Hasenknopf, J.-M. Lehn, G. Baum, D. Fenske, *Proc. Natl. Acad. Sci. USA* **1996**, *93*, 1397–1400.

23 C.-Y. Wong, M. C. W. Chan, N. Zhu, C.-M. Che, *Organometallics* **2004**, *23*, 2263–2272.

24 N. Yoshikawa, A. Ichimura, N. Kanehisa, Y. Kai, H. Takashima, K. Tsukahara, *Acta Cryst.* **2005**, *E61*, 55–56.

25 M. A. Fabre, J. Jaud, J. J. Bonvoisin, *Inorg. Chim. Acta.* **2005**, *358*, 2384–2394.

26 T. Koizumi, K. Tanaka, *Inorg. Chim. Acta.* **2005**, *358*, 1999–2004.

27 M. Abrahamsson, H. Wolpher, O. Johansson, J. Larsson, M. Kritikos, L. Eriksson, P.-O. Norrby, J. Bergquist, L. Sun, B. Åkermark, L. Hammarström, *Inorg. Chem.* **2005**, *44*, 3215–3225.

28 S. Patra, B. Sarkar, S. Ghumaan, M. P. Patil, S. M. Mobin, R. B. Sunoj, W. Kaim, G. K. Lahiri, *Dalton Trans.* **2005**, 1188–1194.

29 H. Hofmeier, E. Herdtweck, U. S. Schubert, *Z. Anorg. Allg. Chem.* **2004**, *630*, 683–688.

30 M. Beley, J. P. Collin, R. Louis, B. Metz, J. P. Sauvage, *J. Am. Chem. Soc.* **1991**, *113*, 8521–8522.

31 M. Maestri, N. Armaroli, V. Balzani, E. C. Constable, A. M. W. C. Thompson, *Inorg. Chem.* **1995**, *34*, 2759–2767.

32 M. Ziegler, V. Monney, H. Stoeckli-Evans, A. von Zelewsky, I. Sasaki, G. Dupic, J.-C. Daran, G. G. A. Balavoine, *J. Chem. Soc., Dalton Trans.* **1999**, 667–676.

33 R. Ziessel, V. Grosshenny, M. Hissler, C. Stroh, *Inorg. Chem.* **2004**, *43*, 4262–4271.

34 V. Grosshenny, R. Ziessel, *J. Organomet. Chem.* **1993**, *453*, C19–C22.

35 D. L. Greene, D. M. P. Mingos, *Transition Met. Chem.* **1991**, *16*, 71–72.

36 E. Alessio, G. Mestroni, G. Nardin, W. M. Attia, M. Calligaris, G. Sava, S. Zorzet, *Inorg. Chem.* **1998**, *27*, 4099–4106.

37 S. Kelch, M. Rehahn, *Macromolecules* **1997**, *30*, 6185–6193.

38 S. Pyo, E. Pérez-Cordero, S. G. Bott, L. Echegoyen, *Inorg. Chem.* **1999**, *38*, 3337–3343.

39 A. Islam, N. Ikeda, K. Nozaki, Y. Okamoto, B. Gholamkhass, A. Yoshimura, T. Ohno, *Coord. Chem. Rev.* **1998**, *171*, 355–363.

40 M. L. Stone, G. A. Crosby, *Chem. Phys. Lett.* **1981**, *79*, 169–173.

41 U. S. Schubert, C. Eschbaumer, P. Andres, H. Hofmeier, C. H. Weidl, E. Herdtweck, E. Dulkeith, A. Morteani, N. E. Hecker, J. Feldmann, *Synth. Met.* **2001**, *121*, 1249–1252.

42 D. M. Klassen, G. A. Crosby, *J. Chem. Phys.* **1968**, *48*, 1853–1858.

43 A. C. Benniston, G. Chapman, A. Harriman, M. Mehrabi, C. A. Sams, *Inorg. Chem.* **2004**, *43*, 4227–4233.

44 A. A. Bhuiyan, J. R. Kincaid, *Inorg. Chem.* **1998**, *37*, 2525–2530.

45 N. Armaroli, *Photochem. Photobiol. Sci.* **2003**, *2*, 73–87.

46 I. M. Dixon, J. P. Collin, J. P. Sauvage, L. Flamigni, S. Encinas, F. Barigelletti, *Chem. Soc. Rev.* **2000**, *29*, 385–391.

47 W. Leslie, R. A. Poole, P. R. Murray, L. J. Yellowlees, A. Beeby, J. A. G. Williams, *Polyhedron* **2004**, *23*, 2769–2777.

48 N. P. Ayala, C. M. Flynn, Jr., L. Sacksteder, J. N. Demas, B. A. DeGraff, *J. Am. Chem. Soc.* **1990**, *112*, 3837–3844.

49 W. Goodall, J. A. G. Williams, *J. Chem. Soc., Dalton Trans.* **2000**, 2893–2895.

50 W. Leslie, A. S. Batsanov, J. A. K. Howard, J. A. G. Williams, *Dalton Trans.* **2004**, 623–631.

51 L. Flamigni, B. Ventura, F. Barigelletti, E. Baranoff, J.-P. Collin, J.-P. Sauvage, *Eur. J. Inorg. Chem.* **2005**, 1312–1318.

52 E. Baranoff, J. P. Collin, L. Flamigni, J. P. Sauvage, *Chem. Soc. Rev.* **2004**, *33*, 147–155.

53 L. M. Vogler, K. J. Brewer, *Inorg. Chem.* **1996**, *35*, 818–824.

54 S. E. Hobert, J. T. Carney, S. D. Cummings, *Inorg. Chim. Acta* **2001**, *318*, 89–96.

55 G. Lowe, A.-S. Droz, J. J. Park, G. W. Weaver, *Bioorg. Chem.* **1999**, *27*, 477–486.

56 K. W. Jennette, S. J. Lippard, G. A. Vassiliades, W. R. Bauer, *Proc. Natl. Acad. Sci. USA* **1974**, *71*, 3839–3843.

57 J. Reedijk, *Proc. Natl. Acad. Sci. USA* **2003**, *100*, 3611–3616.

58 N. J. Wheate, J. G. Collins, *Coord. Chem. Rev.* **2003**, *241*, 133–145.

59 D. R. McMillin, J. J. Moore, *Coord. Chem. Rev.* **2002**, *229*, 113–121.

60 M. Cortes, J. T. Carney, J. D. Oppenheimer, K. E. Downey, S. D. Cummings, *Inorg. Chim. Acta* **2002**, *333*, 148–151.

61 P. Korall, A. Borje, P. O. Norrby, B. Akermark, *Acta Chem. Scand.* **1997**, *51*, 760–766.

62 V. G. Vaidyanathan, B. U. Nair, *Eur. J. Inorg. Chem.* **2004**, 1840–1846.

63 M. Grätzel, *Prog. Photovolt. Res. Appl.* **2000**, *8*, 171–185.

64 J. Husson, M. Beley, G. Kirsch, *Tetrahedron Lett.* **2003**, *44*, 1767–1770.

65 C. Mikel, P. G. Potvin, *Polyhedron* **2002**, *21*, 49–54.

66 M. Beley, C. A. Bignozzi, G. Kirsch, M. Alebbi, J. C. Raboin, *Inorg. Chim. Acta* **2001**, *318*, 197–200.

67 E. Figgemeier, V. Aranyos, E. C. Constable, R. W. Handel, C. E. Housecroft, C. Risinger, A. Hagfeldt, E. Mukhtar, *Inorg. Chem. Commun.* **2004**, *7*, 117–121.

68 E. C. Constable, C. E. Housecroft, M. Neuburger, A. G. Schneider, *Inorg. Chem. Commun.* **2003**, *6*, 912–915.

69 E. C. Constable, R. W. Handel, C. E. Housecroft, A. F. Morales, L. Flamigni, F. Barigelletti, *Dalton Trans.* **2003**, 1220–1222.

70 C. J. Aspley, J. A. G. Williams, *New. J. Chem.* **2001**, *25*, 1136–1147.

71 E. C. Constable, A. M. W. C. Thompson, N. Armaroli, V. Balzani, M. Maestri, *Polyhedron* **1992**, *11*, 2707–2709.

72 K. T. Potts, M. J. Cipullo, P. Ralli, G. Theodoridis, *J. Org. Chem.* **1982**, *47*, 3027–3038.

73 P. Lainé, F. Bedioui, P. Ochsenbein, V. Marvaud, M. Bonin, E. Amouyal, *J. Am. Chem. Soc.* **2002**, *124*, 1364–1377.

74 P. Laine, F. Bedioui, E. Amouyal, V. Albin, F. Berruyer-Penaud, *Chem. Eur. J.* **2002**, *8*, 3162–3176.

75 E. Amouyal, M. Mouallem-Bahout, *J. Chem. Soc., Dalton Trans.* **1992**, 509–513.

76 Y.-Q. Fang, N. J. Taylor, G. S. Hanan, F. Loiseau, R. Passalacqua, S. Campagna, H. Nierengarten, A. van Dorsselaer, *J. Am. Chem. Soc.* **2002**, *124*, 7912–7913.

77 M. I. J. Polson, N. J. Taylor, G. S. Hanan, *Chem. Commun.* **2002**, 1356–1357.

78 C. Metcalfe, S. Spey, H. Adams, J. A. Thomas, *J. Chem. Soc., Dalton Trans.* **2002**, 4732–4739.

79 M. I. J. Polson, E. A. Medlycott, G. S. Hanan, L. Mikelsons, N. J. Taylor, M. Watanabe, Y. Tanaka, F. Loiseau, R. Passalacqua, S. Campagna, *Chem. Eur. J.* **2004**, *10*, 3640–3648.

80 E. C. Constable, D. R. Smith, *Supramol. Chem.* **1994**, *4*, 5–7.

81 G. Albano, V. Balzani, E. C. Constable, M. Maestri, D. R. Smith, *Inorg. Chim. Acta* **1998**, *277*, 225–231.

82 M. Hissler, A. Harriman, A. Khatyr, R. Ziessel, *Chem. Eur. J.* **1999**, *5*, 3366–3381.

83 C. Goze, D. V. Kozlov, F. N. Castellano, J. Suffert, R. Ziessel, *Tetrahedron Lett.* **2003**, *44*, 8713–8716.

84 A. C. Benniston, A. Harriman, D. J. Lawrie, A. Mayeux, K. Rafferty, O. D. Russel, *Dalton Trans.* **2003**, 4762–4769.

85 J. Hjelm, R. W. Handel, A. Hagfeldt, E. C. Constable, C. E. Housecroft, R. J. Forster, *Inorg. Chem.* **2005**, *44*, 1073–1081.

86 K. R. J. Thomas, J. T. Lin, C.-P. Chang, C.-H. Chuen, C.-C. Cheng, *J. Chin. Chem. Soc.* **2002**, *49*, 833–840.

87 R. Passalacqua, F. Loiseau, S. Campagna, Y.-Q. Fang, G. S. Hanan, *Angew. Chem. Int. Ed.* **2003**, *42*, 1608–1611.

88 J. Wang, G. S. Hanan, F. Loiseau, S. Campagna, *Chem. Commun.* **2004**, 2068–2069.

89 N. W. Alcock, P. R. Barker, J. M. Haider, M. J. Hannon, C. L. Painting, Z. Pikramenou, E. A. Plummer, K. Rissanen, P. Saarenketo, *J. Chem. Soc., Dalton Trans.* **2000**, 1447–1462.

90 U. Ziener, E. Breuning, J.-M. Lehn, E. Wegelius, K. Rissanen, G. Baum, D. Fenske, G. Vaughan, *Chem. Eur. J.* **2000**, *6*, 4132–4139.

91 Y.-Z. Hu, G. Zhang, R. P. Thummel, *Org. Lett.* **2003**, *5*, 2251–2253.

92 F. Barigelletti, B. Ventura, J.-P. Collin, R. Kayhanian, P. Gavina, J.-P. Sauvage, *Eur. J. Inorg. Chem.* **2000**, 113–119.

93 O. Johansson, M. Borgstroem, R. Lomoth, M. Palmblad, J. Bergquist, L. Hammarstroem, L. Sun, B. Kermark, *Inorg. Chem.* **2003**, *42*, 2908–2918.

94 R. Dobrawa, F. Würthner, *Chem. Commun.* **2002**, 1878–1879.

95 A. El-Ghayoury, A. P. H. J. Schenning, P. A. van Hal, C. H. Weidl, J. L. J. van Dongen, R. A. J. Janssen, U. S. Schubert, E. W. Meijer, *Thin Solid Films* **2002**, *403–404*, 97–101.

96 J. Otsuki, H. Kameda, S. Tomihira, H. Sakaguchi, T. Takido, *Chem. Lett.* **2002**, *31*, 610–611.

97 T.-Y. Dong, M.-c. Lin, M. Y.-N. Chiang, J.-Y. Wu, *Organometallics* **2004**, *23*, 3921–3930.

98 M. Duati, S. Fanni, J. G. Vos, *Inorg. Chem. Commun.* **2000**, *3*, 68–70.

99 M. Duati, S. Tasca, F. C. Lynch, H. Bohlen, J. G. Vos, S. Stagni, M. D. Ward, *Inorg. Chem.* **2003**, *42*, 8377–8384.

100 M. Poyatos, J. A. Mata, E. Falomir, R. H. Crabtree, E. Peris, *Organometallics* **2003**, *22*, 1110–1114.

101 P. K. Iyer, J. B. Beck, C. Weder, S. J. Rowan, *Chem. Commun.* **2005**, 319–321.

102 M. Shiotsuki, T. Suzuki, K. Iida, Y. Ura, K. Wada, T. Kondo, T. Mitsudo, *Organometallics* **2003**, *22*, 1332–1339.

103 J. D. Holbrey, G. J. T. Tiddy, D. W. Bruce, *J. Chem. Soc., Dalton Trans.* **1995**, 1769–1774.

104 J. Costa, R. Ruloff, L. Burai, L. Helm, A. E. Merbach, *J. Am. Chem. Soc.* **2005**, *127*, 5147–5157.

105 R. Ruloff, G. v. Koten, A. E. Merbach, *Chem. Commun.* **2004**, 842–843.

106 L. F. Szczepura, S. M. Maricich, R. F. See, M. R. Churchill, K. J. Takeuchi, *Inorg. Chem.* **1995**, *34*, 4198–4205.

107 G. Chelucci, A. Saba, D. Vignola, C. Solinas, *Tetrahedron* **2001**, *57*, 1099–1104.

108 E. C. Constable, T. Kulke, M. Neuburger, M. Zehnder, *New. J. Chem.* **1997**, *21*, 1091–1102.

109 A. El-Ghayoury, H. Hofmeier, A. P. H. J. Schenning, U. S. Schubert, *Tetrahedron Lett.* **2004**, *45*, 261–264.

110 H. Hofmeier, A. El-Ghayoury, A. P. H. J. Schenning, U. S. Schubert, *Tetrahedron* **2004**, *60*, 6121–6128.

111 H. S. Chow, E. C. Constable, C. E. Housecroft, M. Neuburger, *Dalton Trans.* **2004**, 4568–4569.

4
Metallo-Supramolecular Terpyridine Architectures[*]

4.1
Introduction

In the preceding chapter, the synthesis and characterization of terpyridine metal complexes were described. The initial focus was on simple *bis*terpyridine ruthenium "[Ru(tpy)$_2$(X)$_2$]" complexes and the tuning of their optical properties; however, compounds containing terpyridines and terpyridine analogs can be used to obtain even more complicated architectures. Thus, by designing tailored ligands, a structural control of the resulting architecture is possible.

This chapter describes recent developments in supramolecular chemistry of [M(tpy)$_2$(X)$_2$] complexes within a single structure. The synthesis and characteristics of extended-supramolecular aggregates, e.g., grid structures, will be considered. It will also be shown how the size of macrocycles possessing one or more *bis*terpyridine metal complex centers can be controlled by the initial polyligand. Furthermore, complexes containing fullerenes, biomolecules, and other exotic building blocks will be presented.

4.2
Dyads and Triads

Many systems containing two or more [M(tpy)$_2$(X)$_2$] units, linked by different spacers, have been synthesized (Figure 4.1). Both homo-dyads (same metal ions) and hetero-dyads (different metal ions) are well known. A variety of such compounds and their properties were reviewed by Sauvage et al. [1] in 1994 and more recently in a couple of tutorial reviews [2, 3].

In general, if an [Ru(tpy)$_2$]$^{2+}$ complex is connected by means of a continuous unsaturated linkage to a similar [Os(tpy)$_2$]$^{2+}$ complex, excitation of the Ru(II) unit can result in energy transfer from Ru to Os. Because of such energy transfer

---

[*] Parts of this chapter are reproduced from *Chem. Soc. Rev.* **2004**, *33*, 373-399 by permission of The Royal Society of Chemistry.

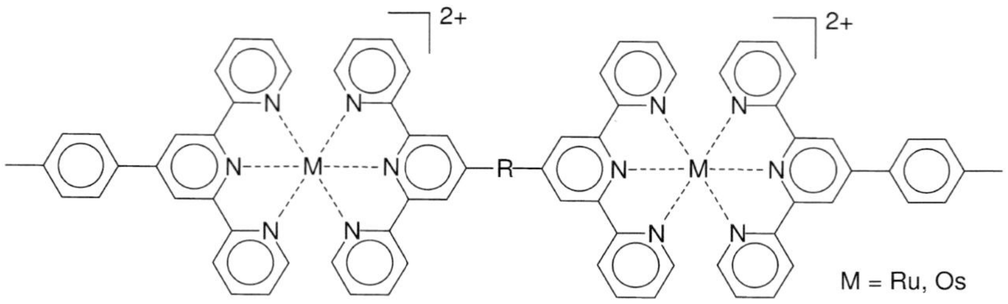

Figure 4.1 Schematic representation of a dyad.

processes (as well as electron transfer processes), these complex arrays constitute interesting examples of potential "molecular wires". Since many dyads were synthesized using rigid *bis*-terpyridine ligands, the resulting complexes have rod-like structures. Representative examples are shown in Figure 4.2. Various heterometallic polyads of bipyridine and terpyridine complexes have been presented in a review by Barigelletti et al. [4].

Starting with a dyad possessing directly connected terpyridines (Figure 4.2a), various types of spacers have been introduced. Intensive work has been performed on a wide range of conjugated spacers of different lengths and compositions; thus, allowing the photophysical properties of the corresponding *bis*-complexes to be fine-tuned.

The groups of A. Harrimann and R. Ziessel have been working for some time on the construction of photoactive wires on the molecular scale [8–16]. In 1996, they reported "stiff" ditopic ligands and oligomers (Figure 4.2c) derived from terpyridines bridged by alkyne spacers comprising one to four ethynyl moieties [8]. These polyalkynyl chains allow very fast electron exchange between photoactive terminals, thus prolonging the triplet lifetimes by a factor of 3000, when compared to the unsubstituted $[Ru(tpy)_2]^{2+}$ complex.

In certain cases, the alkyne bridge undergoes reductive electropolymerization to generate molecular films, which contain metal centers dispersed throughout their conjugated backbone. A major drawback of these rigid bridged systems is that their solubility decreases significantly with the increasing number of acetylene units. To circumvent this problem, Ziessel et al. devised *bis*(terpyridine) ligands bearing ethynylene-phenylene spacers with solubilizing groups in the 2,5-positions (Figure 4.2d) [6, 13], in which, with increasing repeat units, the lowest energy absorption in these compounds is shifted towards lower energy. The same correlation was demonstrated for the emission; for n = 1, the emission wavelength was 434 nm, whereas for n = 5, it was shifted to 472 nm. This behavior showed that, with increasing number of these repeat units, the electronic communication along the molecular axis can be improved.

Reviews of these alkyne-linked dyads [5] and phenylene-linked complexes [1] have appeared. For the *p*-phenylene-bridged dyads (with n = 0–2), it has been shown that conjugation enhanced the luminescence lifetime [7]; however, the phenylene-bridged system showed less luminescence than the directly linked

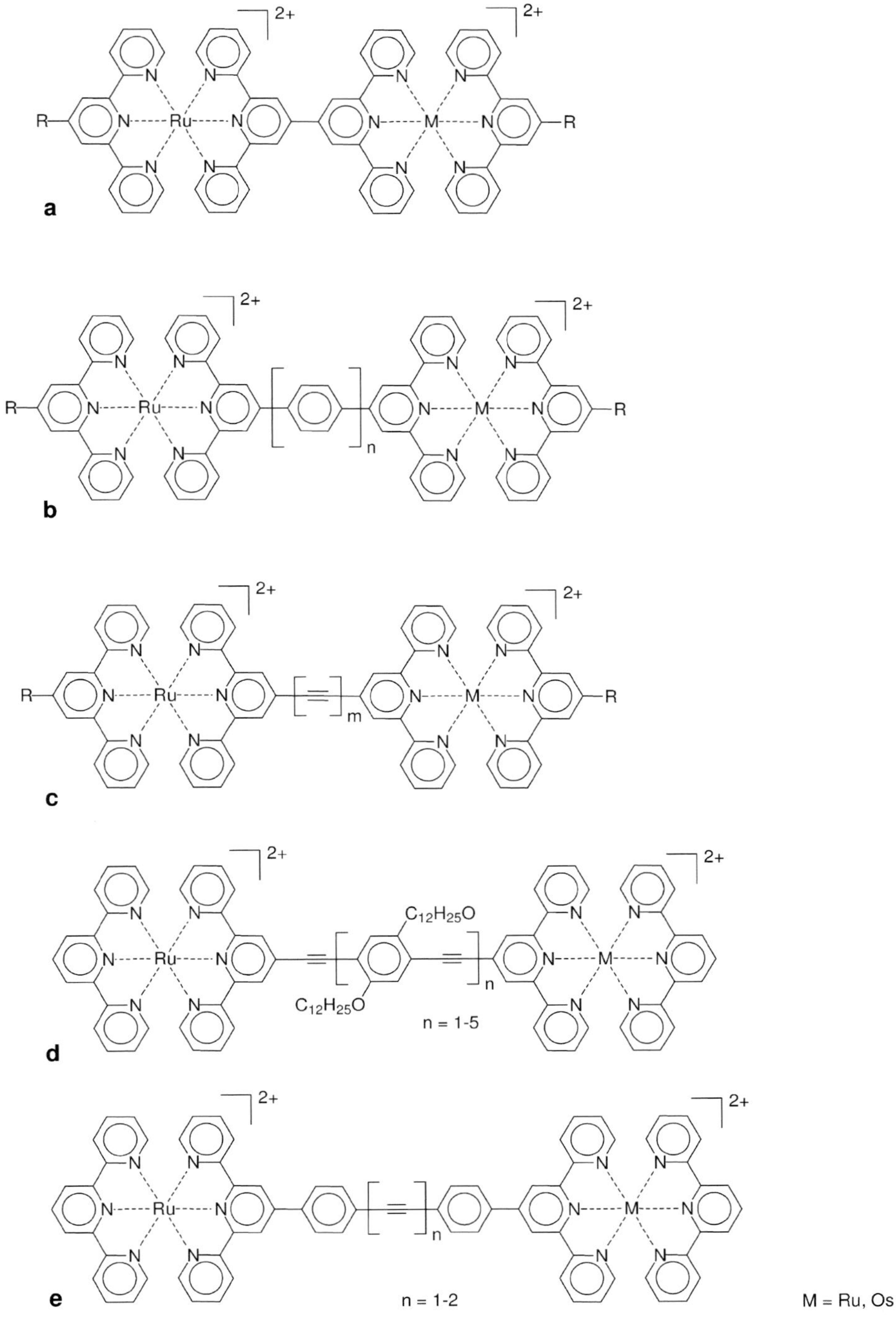

Figure 4.2 Dyads linked by *p*-phenylene and/or ethynyl moieties [1, 5–7].

a	X = Y = C₆H₄Me,	n = 0
b	X = Y = C₆H₄Me,	n = 1
c	X = Y = C₆H₄Me,	n = 2
d	X = SO₂Me, Y = H,	n = 0
e	X = SO₂Me, Y = H,	n = 1
f	X = SO₂Me, Y = C₆H₄Me,	n = 1

Figure 4.3 *p*-Phenylene-linked dyads with donor and acceptor groups [1].

counterparts as a result of the lack of complete delocalization due to the non-planar conformation of the phenyl ring.

Homometallic Ru(II) complexes, as well as Ru(II)-Os(II) dyads, have been prepared using bridging ligands. Synthesis of these systems was performed by first complexing a single coordinating site with a ruthenium fragment, followed by treatment with an osmium precursor as the second step. Such one-dimensional molecular arrays have been prepared and studied in detail by the groups of Sauvage, Barigelletti, and Balzani [17–21]. Energy transfer from the Ru(II) center to that of Os(II) could be observed in the products, even over the extended distance of two phenylene groups.

Further adjustment and fine tuning of the optical properties of this dyad could be achieved by introducing donor or acceptor groups onto the terminal ligands (Figure 4.3) [22]. Electrochemical experiments revealed both metal-metal and ligand-ligand interactions, whereas photophysical studies showed a very efficient energy transfer mechanism, which is most likely based on an electron exchange mechanism.

The architectures of these *bis*-complexes connected by ethynyl groups are interesting due to their enhanced photophysical characteristics. Because of an enhanced delocalization of the electrons in such systems, the triplet energy state is lowered, resulting in efficient luminescence at room temperature.

In a different experiment, Rh(III) was introduced into an analogous series of phenylene-bridged binuclear complexes [23]. When compared to the Ru(II)-Os(II) system, energy transfer from Rh(III) to Ru(II) could be observed, whereas electron transfer from Ru(II) to Rh(III) was found in the case of a direct linkage. Analogous dyads have been prepared using Ru(II) and Co(III) complexes [24], in which electron transfer from Ru(II) to Co(III) was detected.

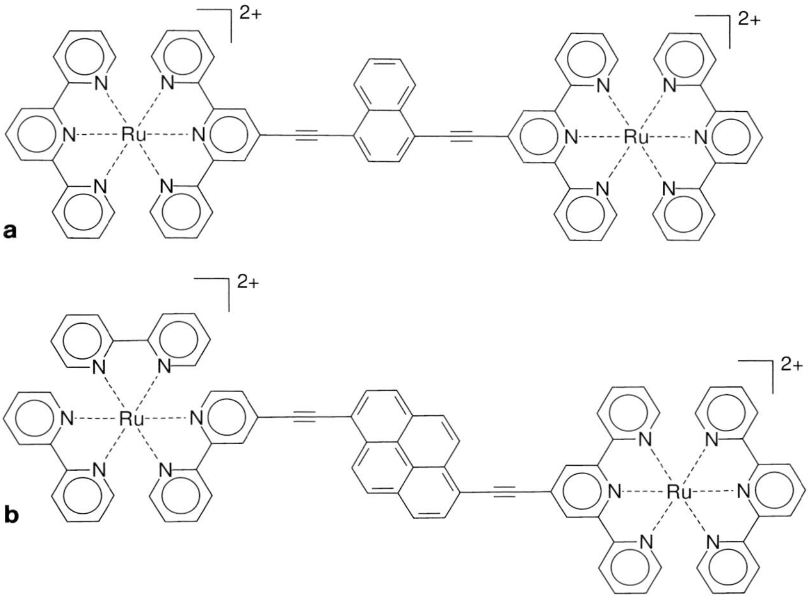

Figure 4.4 Dyads linked by various aromatic groups [25, 26].

In order to study the effect of different central aromatic units on the photophysical properties of these dyads, a 1,4-naphthaleno moiety was introduced into the backbone of the bridging ligand, which was subsequently converted into the ruthenium dyad (Figure 4.4) [25]. Compared to the analogous phenylene-linked dyad, the triplet lifetime is prolonged. Recently, a 1,6-pyreno moiety (Figure 4.4b) that connected an $[Ru(tpy)_2]^{2+}$ complex with an $[Ru(bpy)_3]^{2+}$ complex has been reported [26].

Since the dyads containing the naphthalene-ethyne sequence revealed good photophysical properties, the concept has been extended to other elongated conjugated systems composed of 2 or 4 naphthylethynyl units, which were prepared by Sonogashira coupling reactions of $[Ru(tpy)_2]^{2+}$ complexes bearing one ethynyl group and suitable naphthylethynyl precursors [27]. These compounds, which show room-temperature luminescence, represented another step toward a better understanding of the principles underlining molecular wires (Figure 4.5).

In the dyads discussed so far, rigidity in the spacer is maintained by conjugated systems, such as aromatic or alkyne connections. In order to investigate the effect of conjugation on photophysical and electrochemical properties of these rod-like dinuclear complexes, a saturated bicyclic hydrocarbon was introduced into the linker unit of a Ru(II)-Os(II) dyad (Figure 4.6) [17]. In this system, the electronic interactions were completely decoupled but yet the structural rigidity was maintained. The central bicyclo[2.2.2]octane acts as an insulator, preventing the electronic communication between the metal centers. Therefore, the electro-

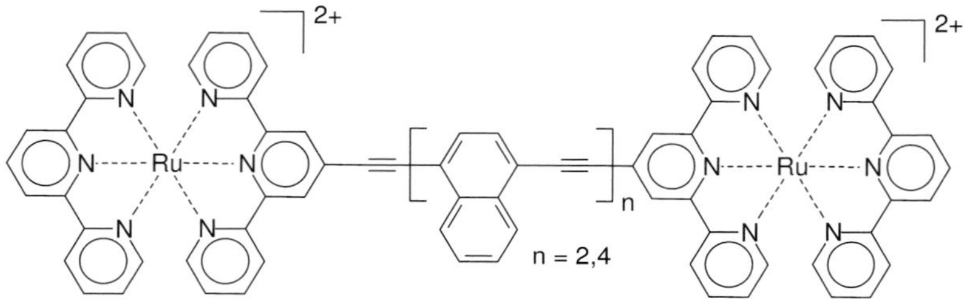

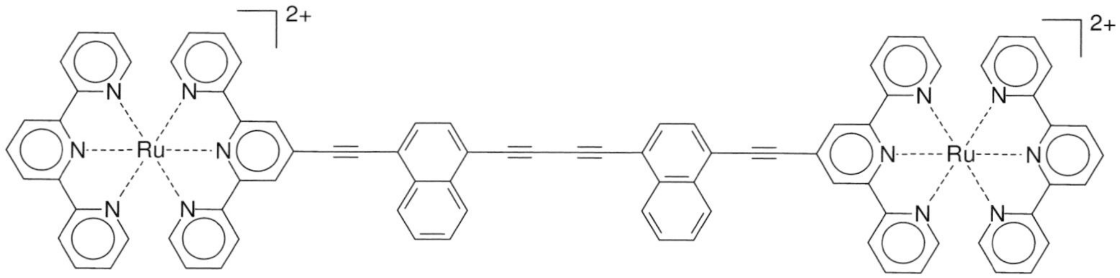

Figure 4.5 Dyads with extended conjugated naphthylethynyl linkers [27].

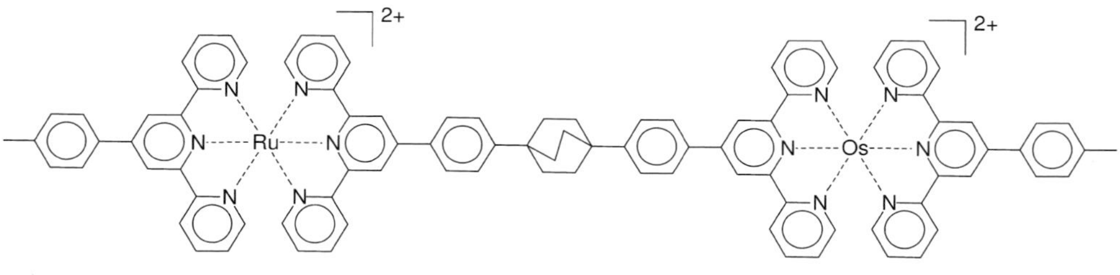

Figure 4.6 Ruthenium-osmium dyad containing an insulating bicyclo[2.2.2]octane connecter [17].

chemistry of such a dyad resembled that of their isolated parent complexes. At room temperature in solution, no energy transfer was found! The lifetime of the Ru-based moiety ($\tau = 1.1$ ns) is too short to allow energy or electron transfer; whereas at low temperatures (77 K) in a rigid matrix, the triplet lifetime was long enough ($\tau = 10.5$ µs) to allow an energy transfer with a rate constant of 4.4×10^6 s^{-1}. As expected, no electron transfer was detected due to the isolating linker.

The electronic coupling/decoupling of the metal centers in these dyads could be controlled by the degree-of-conjugation of the central biphenyl linker, which was structurally locked by the length of the bridging "strap" (Figure 4.7) [28, 29]. The synthesis and photophysical characteristics of a related linear terpyridine-based trinuclear Ru(II)-Os(II) array, composed of two [Ru(tpy)$_2$]$^{2+}$ termini connected by means of an alkoxy-strapped biphenylene units to a central [Os(tpy)$_2$]$^{2+}$ core, have been described [30].

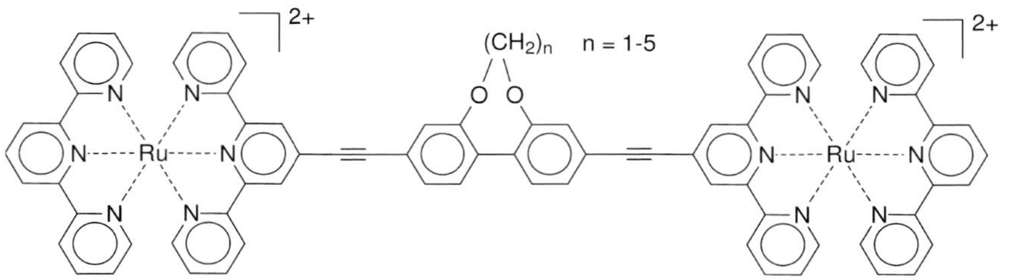

Figure 4.7 Control of the dyad communication as a factor of the connecting cyclic "strap" [28].

Dyads containing heteroaromatics (bipyridine, phenanthroline, 2,2'-bipyrimidine, as well as ferrocene [31–33]) in the linker have also been reported; however, their preparation by reacting the *bis*-terpyridine ligand with terpyridine-ruthenium(III) fragments was unsuccessful [34]. An electroactive luminescent switch prepared from [Ru(tpy')(tpy'')]$^{2+}$, where tpy' is a hydroquinone-moiety and tpy'' is a 4'-phenylethynyl fragment has recently appeared [35]. To circumvent this assembly problem, the compounds were prepared by Pd-catalyzed cross-coupling of preformed complexes of 4'-bromoterpyridine with the corresponding diethynyl species [36] (Scheme 4.1).

*Bis*terpyridinyl transition metal oligomers with up to three metal centers have been reported by Colbran et al. [37]. Coupling the aniline group in the heteroleptic [Ru(4'-(4-H$_2$NC$_6$H$_4$)tpy)$_2$]$^{2+}$ [38] complex with 4'-(4-chlorocarboxyphenyl)terpyridine yielded an amido-linked binucleating ligand with a coordinated and a non-coordinated terpyridine domain (Scheme 4.2). Using this ligand, dimers were readily accessible. These authors used Co(II), Fe(II), and Ru(II) ions to connect the ligands. The application of NMR was of particular interest for the Ru(II)-Co(II)-Ru(II) triad, in which the paramagnetic ion leads to a downfield shift of the signals of the terpyridine hydrogens by ca. 15 to 90 ppm (Knight shift, see also Chapter 3).

On the way to molecular devices, a specific dyad was designed that could act as a T-junction relay (molecular switch) [39]. Thus, a spiropyran moiety that is incorporated into the linker unit underwent a light-activated ring-opening to the merocyanine form and then thermally recyclized. In the closed form, the delocalized bridge could act as a "molecular wire" (it is known that the MLCT state involves the LUMO, which spans across both terminals), whereas, in the merocyanine form, the alternative pathway to the side group is opened (Scheme 4.3).

All multinuclear complexes herein reported were composed of 4'-functionalized terpyridines giving rise to rod-like complex arrays. In contrast to these approaches, the bridging ligand depicted in Figure 4.8 consists of three terpyridines, which are linked in the 4- and 4''-positions [40]. Different structures are, therefore, accessible [41]. Compared to the *tris*-ligands described before, in this particular case, individual terpyridine moieties are linked in the 5- and 5''-positions by ethyl and ethylene groups, respectively. Trinuclear Ru(II) complexes with unfunctionalized terpyridine revealed luminescence at room temperature for the ethylene-linked complex and no emission for the alkyl-spacer-separated triad.

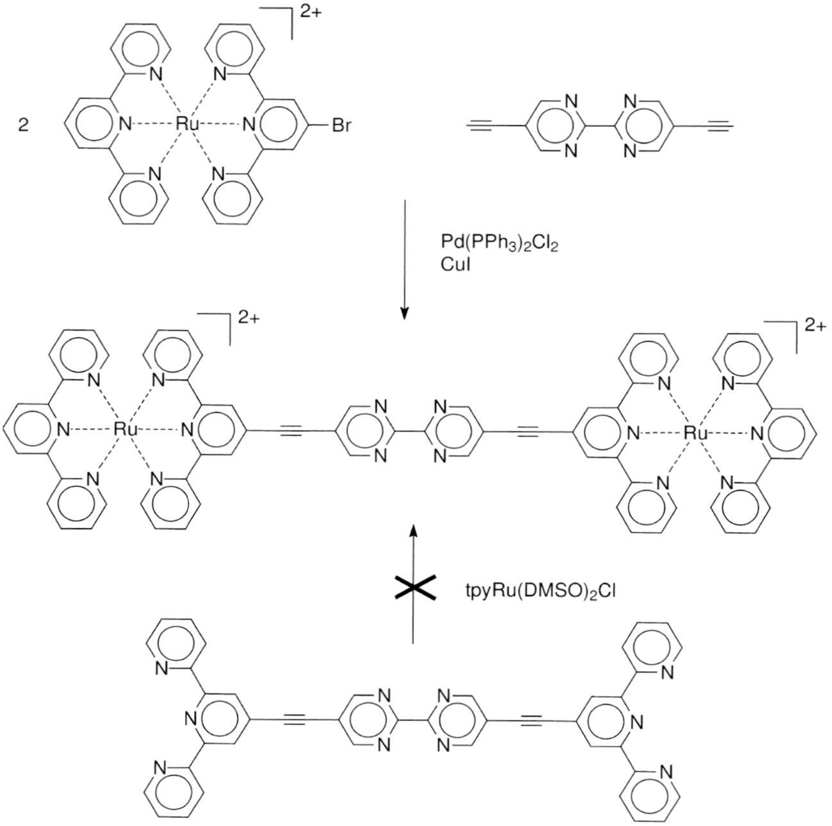

Scheme 4.1 Preparation of a dyad with a bipyrimidine linker
(the same protocol was applied to an analogous phenanthroline system) [34].

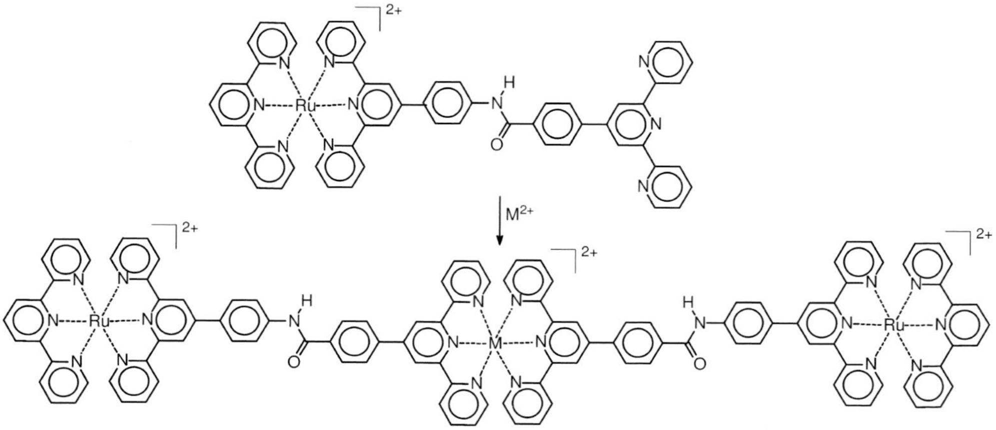

Scheme 4.2 Linear molecules with three metal centers prepared by Colbran and Storrier [37].

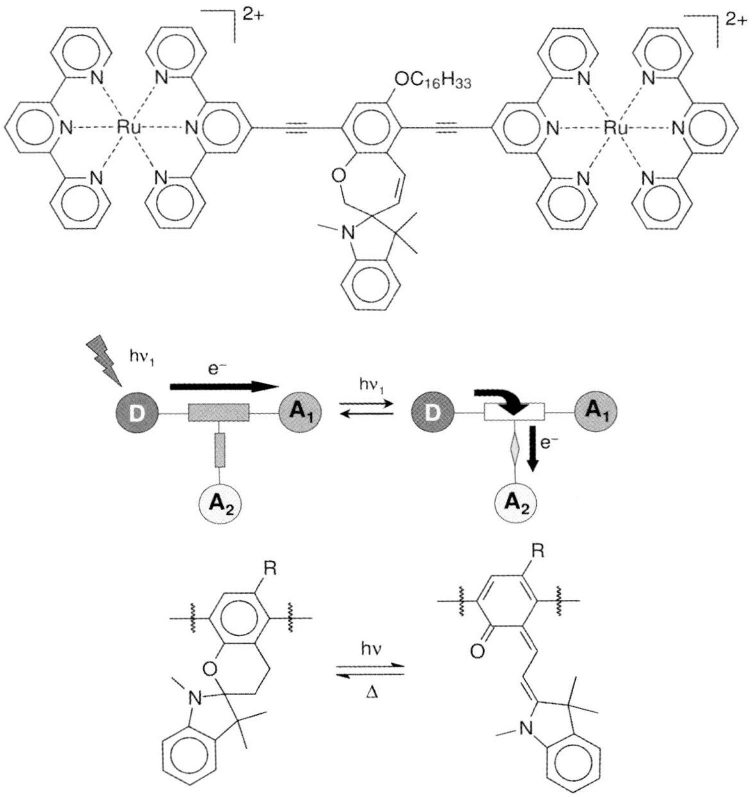

Scheme 4.3 Dyad containing a spiropyrane that can act as a molecular switch (T-junction relay) and schematic representation of the switching process [39].

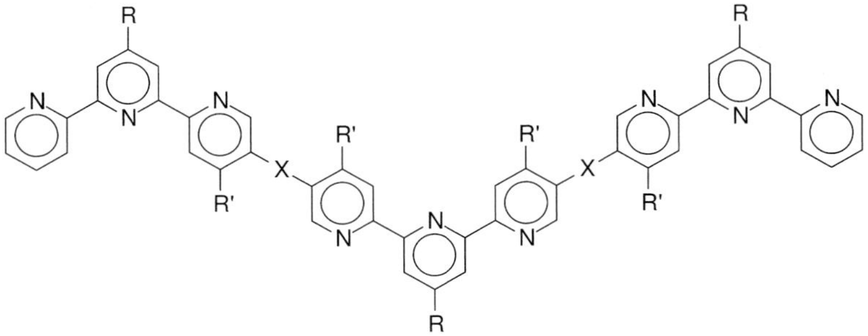

Figure 4.8 Ligand for a 5,5″-linked triad [41].

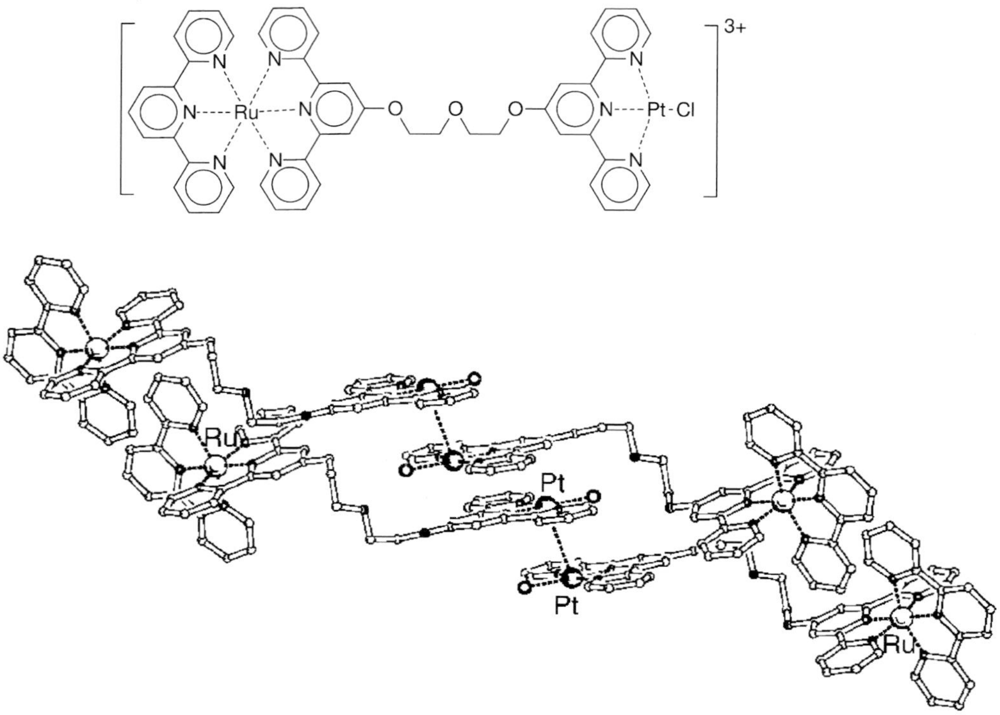

Figure 4.9 Schematic representation and crystal structure of the Ru(II)-Pt(II) dyad. (Reprinted with permission from [45]).

In view of the possibility of anticancer reagents (terpyridine platinum complexes [42, 43] and also certain ruthenium species [44] showed anti-carcinoma activity), a flexible Ru(II)-Pt(II) dyad was designed (Figure 4.9) [45]. The crystal structure revealed a stacking of the mono-terpyridine Pt(II) units, in which the two platinum centers paired off, as shown by the short interatomic distance.

Closely related to the chiral terpyridine systems presented earlier, chiral multinuclear complexes have also been reported by Abruña et al. in which two "dipineno"-(5,6:5",6")-fused 2,2':6',2"-terpyridine ligands were linked in the 4'-position by a biphenyl moiety (Figure 4.10a) [46]. A slightly different approach consisted of either the dipineno-groups or a p-xylene linker (Figure 4.10b). Both ligand systems were used to obtain dinuclear Ru(II) complexes and coordination polymers that exhibit room-temperature luminescence due to the aromatic connector. The ligands were also applied in the preparation of Fe(II) coordination polymers (see also Chapter 5) [47, 48].

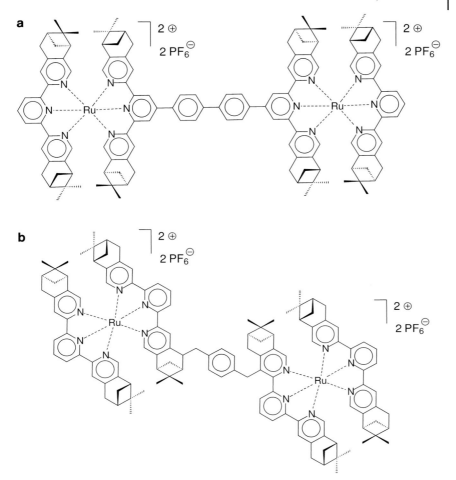

Figure 4.10 Chiral Ru(II)-Ru(II) dyads [46].

4.2.1
Switchable Dyads and Triads

Multinuclear $[M(tpy)_2]^{2+}$ complexes, especially those capable of energy and/or electron transfer, are promising materials for the future in optical nano-devices or solar cell applications. In the field of devices, molecular switches whereby electrochemical or optical activities can be switched simply and reversibly, are of special importance.

The *mono*-$[Ru(tpy)_2]^{2+}$ complex with an appended free ligand (separated by zero to two phenylene rings as described above) can not only be complexed, but can also possess other interesting properties, e.g., upon protonation of the free terpyridine moiety, luminescence properties can be modulated in a reversible

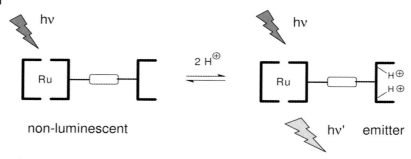

Scheme 4.4 Luminescence switching by protonation of a free terpyridine moiety [49].

manner. Thus, after protonation initiated by a pH change, the non-luminescent complex became luminescent, giving access to a form of "switchable" luminescence (Scheme 4.4) [49]. Moreover, complexes of this type can be further extended through complexation by addition of different metal ions, i.e. Fe(II) or Zn(II), to result in ABA-triads (Scheme 4.5). The almost non-luminescent $[Ru(tpy)_2]^{2+}$ chromophore, functionalized with a free terpyridine fragment, was complexed with Zn(II) ions, leading to a luminescent rod-like complex array revealing a luminescence enhancement factor (EF) larger than 10 [22]. Because of the reversible, relatively weakly coordinated zinc complex, this system also gives rise to "switchable" emitters.

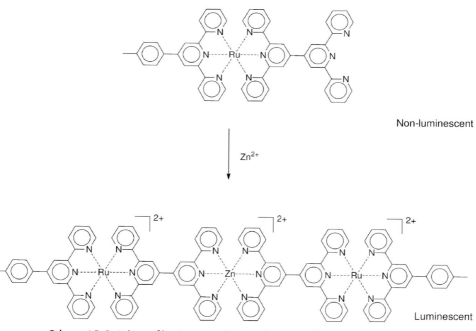

Scheme 4.5 Switching of luminescence by complexation with Zn(II) ions [22].

A different approach to tunable metal complex arrays includes the incorporation of a different complexing moiety, such as bipyridine, within the bridging ligand. One of the first examples was the dyad, bridged by a 2,2′-bipyridine-5,5′-diethynyl group [5, 12], in which the bridging ligand was complexed with various metal ions. By this method, the luminescence properties of the parent complex can be tuned; therefore, weakly binding ions caused an increase in the luminescence of the $[Ru(tpy)_2]^{2+}$ units, whereas strong binders like Ag^+ or Hg^{2+} decreased the luminescence. The rationale for this behavior was attributed to the reducibility of these bridging metal centers making a photoinduced electron transfer possible, which, in turn, resulted in luminescence quenching. When the nitrogen atoms were methylated to afford a viologen-moiety, luminescence quenching through electron transfer processes resulted [5, 12].

Loiseau et al. [50] recently reported a dyad in which the $[Ru(tpy)_2]^{2+}$ termini were directly attached to a bridging bipyridine moiety (see Figure 4.11). An enhancement of luminescence was found when the bipyridine was protonated by addition of acids as well as after complexation with Zn(II). Utilizing cyclic voltammetry, it was found that the electronic interaction between peripheral chromophores was enhanced by zinc coordination. Other related expanded bidentate 2,2′-bipyridine ligands linked to one or two terpyridine units have been synthesized and converted to the $[Ru(bpy)_3]^{2+}$ complexes by treatment with $[Ru(bpy)_2(Cl)_2]$ [51].

A promising approach towards switchable luminescent chromophores was demonstrated by the azo-linked Ru(II)-Os(II) dyads [52]; the related $[Ru(tpy)_2]^{2+}$ and $[Os(tpy)_2]^{2+}$ homo-complexes were also synthesized (Scheme 4.6). In these systems, energy transfer can be switched/triggered by a redox reaction. In its neutral state, the azo-group quenches the luminescence of both Ru(II) and Os(II); however, reduction of the azo-linker afforded a luminescent homodinuclear Os(II) complex at room temperature and the corresponding homodinuclear Ru(II) complex showed luminescence at 77 K. In the heterodinuclear complex, an energy transfer from Ru(II) to Os(II) occurred, resulting in a strong osmium luminescence (Scheme 4.6a).

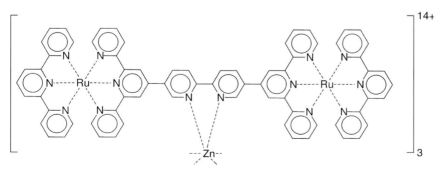

Figure 4.11 Switchable dyad containing a central bipyridine fragment in the bridging ligand [50].

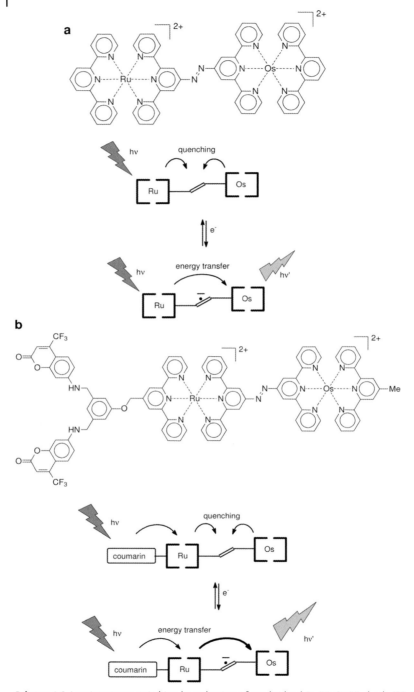

Scheme 4.6 Luminescence switching by reduction of azo-bridged Ru(II)-Os(II) dyads [52, 53].

By attaching a coumarin group, the efficiency could be further improved [53], i.e. energy transfer from the coumarin via the Ru(II) complex to Os(II) was observed (Scheme 4.6b). While the simple $[Ru(tpy)_2]^{2+}$-azo-$[Os(tpy)_2]^{2+}$ dyad revealed an energy transfer of 40%, the coumarin-$[Ru(tpy)_2]^{2+}$-azo-$[Os(tpy)_2]^{2+}$ triad showed an efficiency of more than 70%. Redox-stimulated switching could be performed efficiently. Besides phenyl and pyridinyl groups, other heteroaromatic groups have also been introduced into multinuclear complex arrays of which a prominent example of dyads and triads is the "family" possessing thiophene(s) (Figure 4.12).

Besides model *mono*nuclear complexes bearing one or two thiophenes, dyads and triads with Ru(II) were prepared successfully [54, 55]. While the mononuclear complexes already showed an enhanced emission, the multinuclear complexes revealed a luminescence very similar to that of the bipyridine Ru(II) complexes. This behavior can be explained by the stabilization of a cluster of luminescent ^{3}MLCT levels in the array by intermetal electronic communication of the complex units leading to a higher energy gap and hence to a lower probability of non-radiative deactivation.

Consequently, this approach has been extended to dyads and triads where the Ru(II) complexes were combined with those of Os(II) in order to study energy transfer and electron transfer processes [56]. The mixed complexes were also luminescent and Ru(II) $\rightarrow$ Os(II) energy transfer was observed. As a result, luminescence was enhanced when compared to the *homo*metallic complexes; a five-fold enhancement was also found for the related Ru(II)-Os(II)-Ru(II) triad.

The well-known chemistry of coupling alkynes to terpyridine has also been applied to attach terpyridine moieties to thiophenes [57]. In this way, systems where the terpyridine and thiophene moieties are separated by an alkyne were generated. *Homo*metallic ruthenium dyads as well as an ABA-triad [Ru(II)-Zn(II)-Ru(II)] were prepared. The thiophene unit was found to act as an insulator, thus preventing full delocalization. An enhanced luminescence was found when compared with that of either the uncomplexed parent compounds or phenylene analogs due to an improved stabilization of the triplet state. The thiophene-linked multinuclear complexes may eventually lead to applications such as molecular wires or light-harvesting devices.

Another step toward molecular wires was described recently using *bis*-terpyridinyl ligands [58, 59] separated by spacers with up to five ethynyl-thiophenyl units (Figure 4.13) [60]. Compared to the dyads with a single thiophene unit, the extended complexes showed weaker luminescence, probably due to a decreased electronic communication of the $[Ru(tpy)_2]^{2+}$ complexes or to an interplay of the close-lying excited levels.

Bipyridine ruthenium(II) complexes have also been employed, in a dyad with $[Os(tpy)_2]^{2+}$ complexes, as a photosensitizer for enhanced luminescence [61]. Construction of the $[Ru(tpy)_2]^{2+}$-($\equiv$)-$[Ru(tpy)_2]^{2+}$ and $[Os(tpy)_2]^{2+}$-($\equiv$)-$[Ru(bpy)_3]^{2+}$ as well as the novel photoactive properties of these hybrid complexes have been reported [15]. This system has been made switchable by the introduction of an aminodipyrido[3,2-*a*:2′,3′-*c*]phenazine moiety as a linker (Figure 4.14a). The

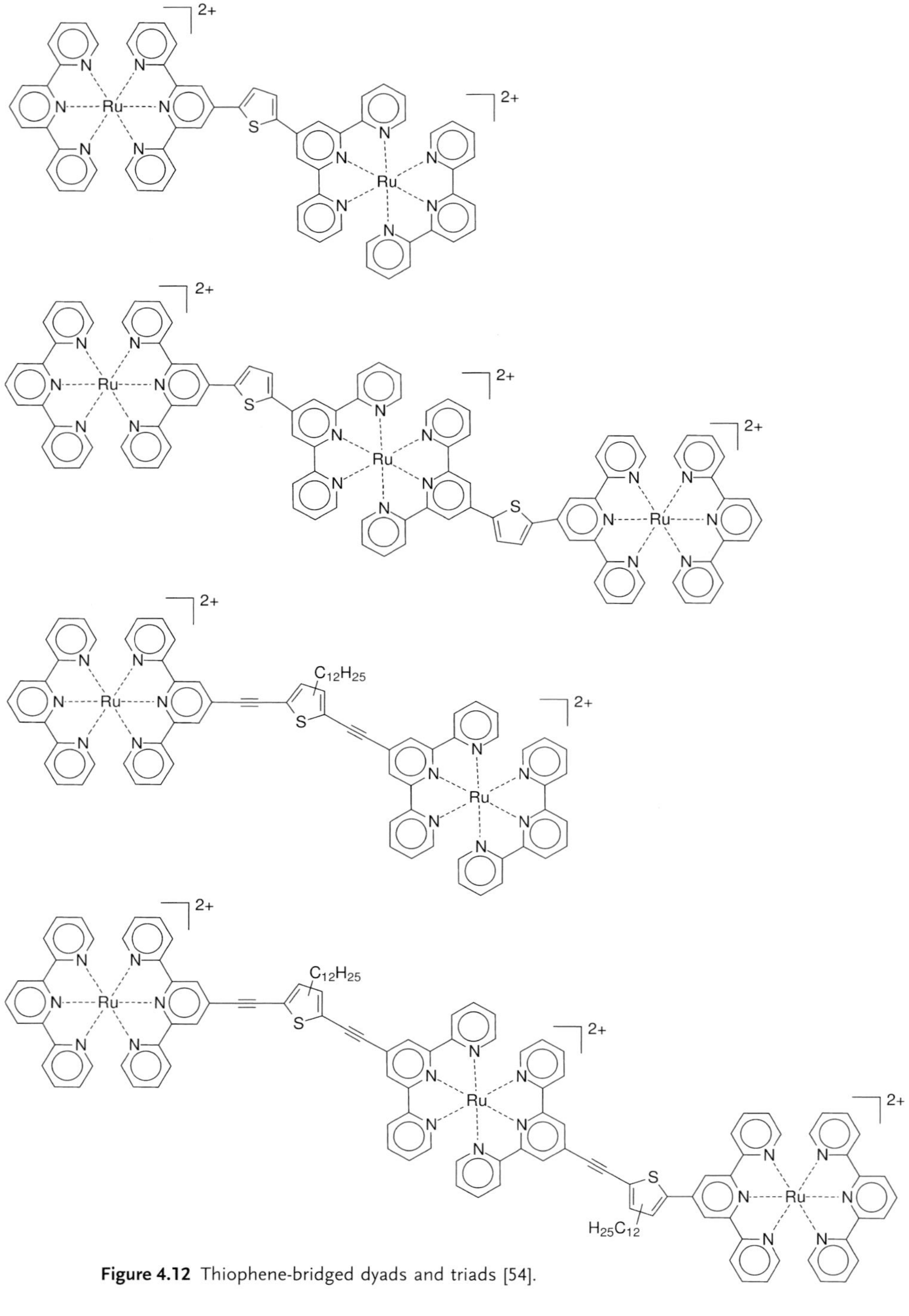

Figure 4.12 Thiophene-bridged dyads and triads [54].

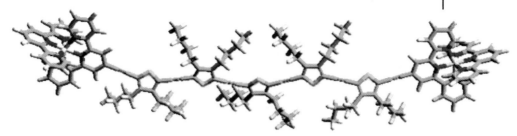

Figure 4.13 Structure (molecular modeling) of a dyad bridged with five di-3,4-substituted thiopheneethynyl units.
(Reprinted with permission from [60], © 2004 American Chemical Society).

a

a R = H
b R = ph

b

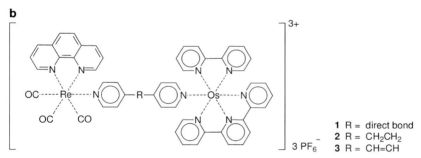

1 R = direct bond
2 R = CH$_2$CH$_2$
3 R = CH=CH

Figure 4.14 [Ru(phen)(bpy)$_2$]$^{2+}$ -R-[Os(tpy)$_2$]$^{2+}$ (a) [61] and [Re(phen)(py)(CO)$_3$]$^{1+}$ -R-[Os(tpy)(bpy)(py)]$^{2+}$ (b) dyads [62].

osmium emission could even be increased by the application of 4,4′-diphenyl-2,2′-bipyridine instead of single bipyridine. "Switching" was easily performed simply by protonation of the connecting unit. The energy transfer system, containing an osmium center as the receiving unit, was extended to a "hetero" system consisting of a novel [Os(tpy)(bpy)(py)]$^{2+}$ complex (Figure 4.14b) to which a [Re(phen)(py)(CO)$_3$]$^{1+}$ complex was connected via an ethylene and a vinylene group; an efficient energy transfer was observed [62].

4.3
Supramolecular Assemblies

Thus far, the oligonuclear complexes that have been considered possessed two separated *bis*terpyridine metal complexes linked by spacers of various types and lengths. Another possibility for constructing extended complex architectures is when two or more terpyridine motives are combined with shared nitrogen atoms (Figure 4.15).

Because these types of ligands have fixed, well-defined geometries, they have been used as building blocks for the construction of extended supramolecular architectures. One ligand example is *tetra*-2-pyridyl-1,4-pyrazine (tppz), which can best be envisioned as two fused terpyridines sharing the central aromatic ring. Such an orientation, with its rigid composition, leads to the construction of rod-like structures (Figure 4.16a) [1]; whereas, the *mono*metallic compounds and the spacer-separated polyads show luminescence. The multimetallic complexes do not emit at room temperature because of metal-metal interaction.

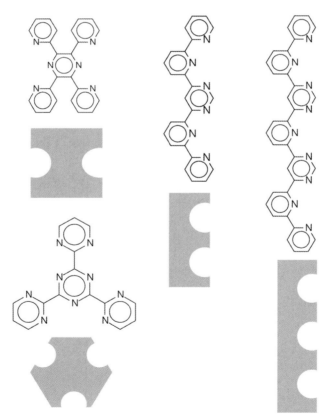

Figure 4.15 Building blocks for supramolecular architectures.

Pyrazine ligands have also been used to synthesize mixed Ru(II)/Rh(III) complexes (Figure 4.16b) [63], in which the *mono*metallic tppz-tpy Ru(II) complex was treated with $RhCl_3$. In that the rhodium complex is known to be an electron acceptor, quenching of luminescence occurred with 80% efficiency via an electron transfer process.

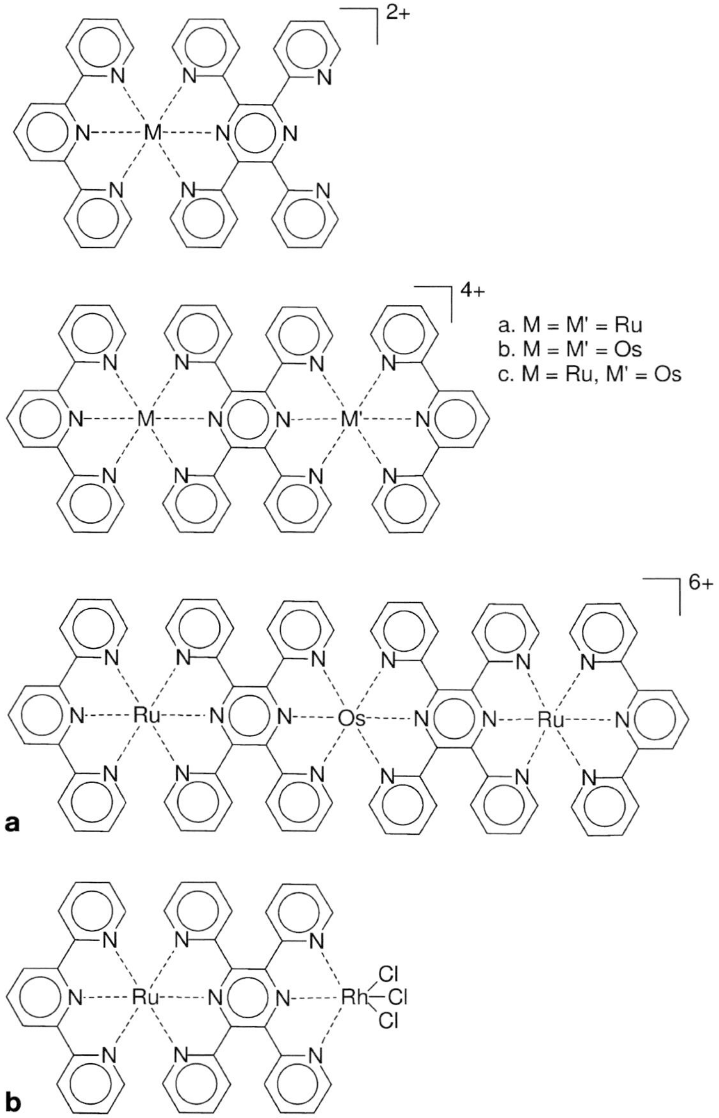
2+
4+
a. M = M' = Ru
b. M = M' = Os
c. M = Ru, M' = Os
6+
a
b
Ru
Os
Rh
Cl

Figure 4.16 Complexes derived from the pyrazine ligand [1, 63].

4.3.1
Grids and Racks

The pyrazine ligand (Figure 4.15a) is composed of two fused terpyridines, pointing in *opposite* directions, allowing the formation of rod-like assemblies. The pyridine rings can also be combined with pyrimidine rings, resulting in ligand systems with fused terpyridine subunits (Figure 4.15b) pointing in the *same* direction. Two different kinds of assemblies can be obtained from these ligands, namely "grids" and "racks". In its free uncomplexed form, the molecule has a helical shape [64, 65], because only this form allows the favored *trans*-conformation of the nitrogen atoms; therefore, the ligand has to unfold to form complexes.

This type of ligand was first reported by *Lehn et al.* in 1995 and was prepared by means of a Stille-type cross-coupling of 6-stannylated bipyridine with 4,6-dichloropyrimidines [66]. Besides the unsubstituted ligand, a ligand with a central 9-anthryl substituent incorporated in the 5-position of the pyrimidine ring has been synthesized (Figure 4.17a). Di- and trinuclear rack-like structures (Figure 4.17b and c) were achieved by end-capping the terpyridine subunits with [Ru(tpy)(Cl)$_3$] under reductive conditions.

Crystal structure analysis revealed that the ligand axis is bent because of the less-than-perfect octahedral geometry for each complex center. This bend was reduced in the case of the anthracene-containing complexes because of the steric repulsion between this central aryl moiety and juxtaposed terpyridine units (Figure 4.18); the anthryl group is sandwiched between the neighboring terpyridines. This example illustrates an internal substitution pattern; the terminal substitution has been demonstrated [67] in an attempt to address the generation of Borromean rings (see below).

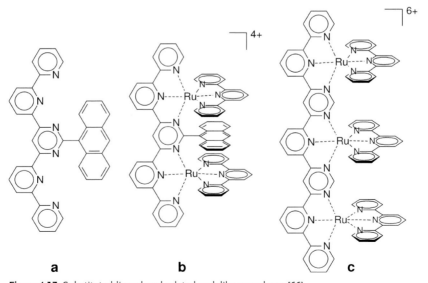

Figure 4.17 Substituted ligand and related rack-like complexes [66].

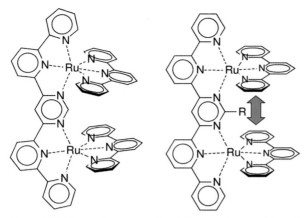

Figure 4.18 Distorted geometry of rack-like complexes and its reduction by a bulky substituent in the 5-position of the pyrimidine ring.

The green coloration for these complexes was explained by the nature of the bridging ligands and metal-metal interactions. UV-vis spectroscopy showed three MLCT absorption bands due to splitting of the π^* level. Different substituents, such as methyl and phenyl groups, have been introduced into the 5-position of the pyrimidine leading to different bending angles for the bridging ligand and convergence angles of the terpyridines; a simple methyl group was found to be the ideal option to align the geometry [68]. In that same publication, *mono*nuclear complexes with only one bridging ligand coordinated to the metal ion were reported; however, the synthesis was more challenging because of activation of the second donor site after initial coordination.

Photophysical experiments revealed an emission originating from the anthracene-containing rack complex in the infrared region [69]. Comparison of the luminescence spectra and lifetimes with those of the luminescence properties of the subunits showed that, in these supramolecular species, excitation energy flows with unitary efficiency to the lowest excited state regardless of which chromophoric subunit was excited. Luminescence could be observed at room temperature as well as at low temperatures in a rigid matrix, and the absorption as well as luminescence properties could be fine-tuned by changing the substituent at the pyrimidine ring. Energy transfer from the central complexes to the peripheral ones could also be observed [70].

The first examples of grid-like assemblies ([2 × 2] grid; Scheme 4.7a), based on bipyridine subunits complexed by Ag(I) or Cu(I) ions, were reported in 1992 by Youinou et al. [71] In 1994, Lehn et al. [72] devised the larger [3 × 3] grid (Scheme 4.7b). Most of the work up to now on these architectures has been performed in the Lehn laboratories.

The first publications on grids from *tris*dentate ligands appeared in 1997 [64, 65]. Through addition of an equimolar amount of hexacoordinating metal ions to *bis*coordinating ligands, analogous coordination arrays are obtained by self-

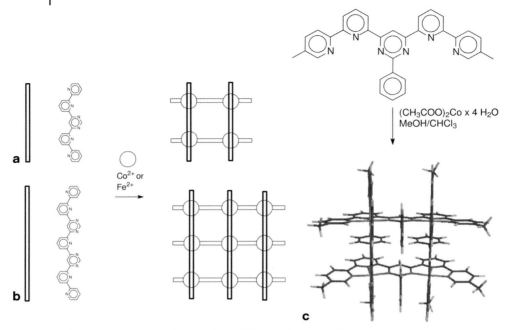

Scheme 4.7 [2 × 2] (a) and [3 × 3] (b) grid-like complexes as well as the Schubert ligand (c) [64, 65].

assembly. Supramolecular grids have also been synthesized by Schubert et al. [73] using 4,6-*bis*(5″-methyl-2″,2′-bipyridin-6′-yl)-2-phenylpyrimidine (Scheme 4.7c), which was synthesized using a Stille-type coupling in a two-step procedure from 2,6-*bis*(trimethyltin)pyridine (via a stannylated bipyridine) and 4,6-dichloro-2-phenylpyrimidine. This ligand has been prepared by an improved procedure in which sodium stannane was applied, avoiding the original problematic lithiation step. This polyligand, when treated with $Co(OAc)_2$, assembled into a [2 × 2] grid, which was successfully detected in an unfragmented form by MALDI-TOF-MS [74]. An antiferromagnetic spin coupling could be found in the cobalt grid [75]. In the case of Fe(II) grids, a spin-crossover from high to low spin could be triggered by temperature, pressure, and light [76].

In order to extend the grid-like structures, the synthetic approach was expanded from *bis*coordinating to *tris*coordinating ligands, which gives entrance to the larger [3 × 3] grids (Scheme 4.8) [77]. Depending on the particular ligand, metal ion(s), and reaction conditions, different grid architectures were formed. The *tris*terdentate ligand composed of three terpyridine subunits was shown to form either the complete [3 × 3] or incomplete [3 × 2] grids. $Zn(BF_4)_2$ with ligand **a** (Scheme 4.8) and Hg(II) triflate with ligand **b** lead to [3 × 3] structures, whereas the other ligands or metal ions, such as Co(II), formed incomplete [3 × 2] grids. Crystal structure analysis of this [3 × 2] grid revealed the central non-coordinating complex moieties

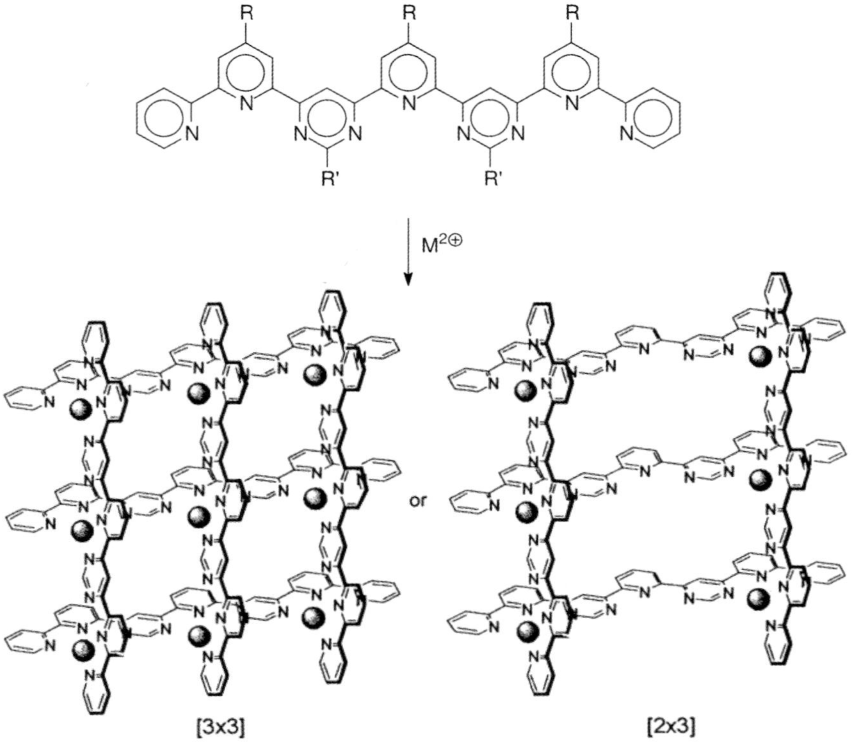

Scheme 4.8 [3 × 3] and [2 × 3] grids (**a**: R = R′ = H, **b**: R = H, R′ = SnPr, **c**: R = Me, R′ = H, **d**: R = Ph, R′ = H). (Reprinted with permission from [77]).

to be in a transoid conformation, which is more stable than the cisoid one because of steric hindrance of the ligands as well as influence of the metal ions and/or counterions used. Pb(II) ions were found to be the most suitable metal ions for the assembly of extended grid structures because their larger size causes less ligand distortion.

Using this metal, Lehn et al. have generated some very large structures [78]; for example, the *tetra*coordinating ligand was assembled into [4 × 4] grids with Pb(II) ions. By simply adjusting the stoichiometry, different structures could be obtained, as shown in Figure 4.19 [79]. The complete [4 × 4] grid was converted into a double-cross-shaped structure via a double-T shaped [2 × 2] grid structure through addition of lead triflate to a solution of the ligand.

Besides X-ray structure analyses, NMR has been shown to be a suitable tool for the investigation of such complexes. The structure of the [4 × 4] grid has been determined by ^{207}Pb NMR (Figure 4.20), which revealed four different metal centers; moreover, the existence of all types of grids up to the [4 × 4] systems has been shown by electrospray MS [80].

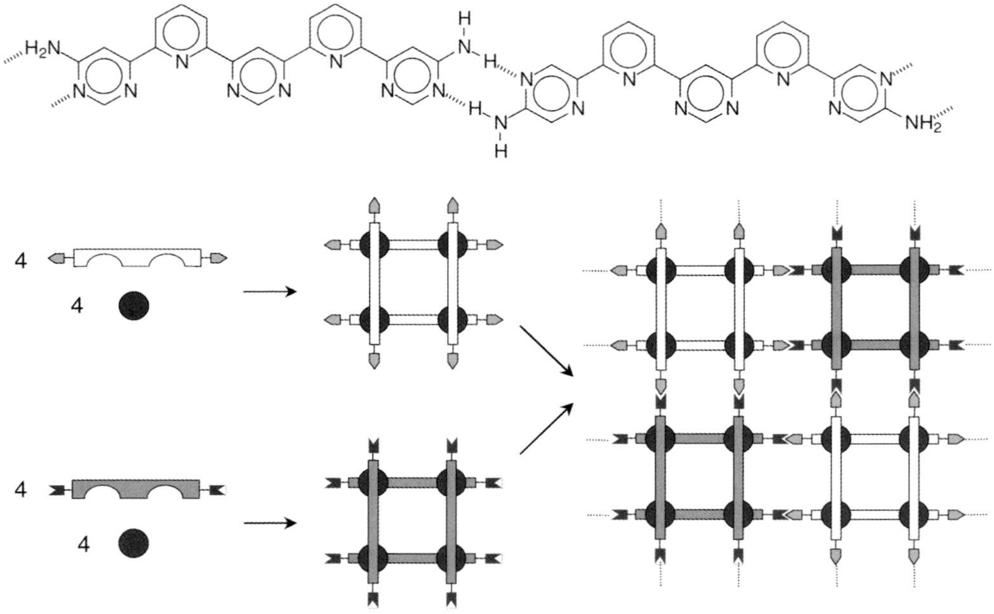

Scheme 4.10 Two-step self-assembly to a "hypergrid" by *H*-bonding [84].

self-assembled tetranuclear [2 × 2] grids in which the nitrate ion was encapsulated, as supported by X-ray crystal structures [83].

In an extension of the self-assembly principle, a two-level self-organization can be achieved by the use of two different ligands that bear complementary *H*-bonding units at the periphery (Scheme 4.10) [84]. These compounds were converted to highly organized grid-like complexes ["grid-of-grids"] by mixing stoichiometric amounts of each complex [85].

Complexes composed of different metal centers could be assembled in this fashion, giving rise to a novel chessboard-like motif. Another approach consisted of introducing pyridine rings in the 4′-position of the terpyridine subunits [86]. The free ligand, as well as the resulting grid complexes, could be assembled on graphite surfaces. Weak CH-N *H*-bonding interactions were also postulated by Lehn et al. based on STM observations and comparisons with molecular modeling results.

Besides racks and grids, other architectures are also conceivable. When Pb(II) ions were added to a mixture of *tris*-2,4,6-(2-pyrimidyl)-1,3,5-triazine (ligands consisting of two or three fused terpyridine moieties), self-assembly took place, resulting in cylindrical cage-like complexes constructed of either 6 or 9 metal centers, respectively (Scheme 4.11) [87]. In this case, 36 (complex **a**) or 54 (complex **b**) coordination bonds between 11 (**a**) or 15 (**b**) components, respectively, are formed in one reaction step, demonstrating the power and selectiveness of the self-assembly process.

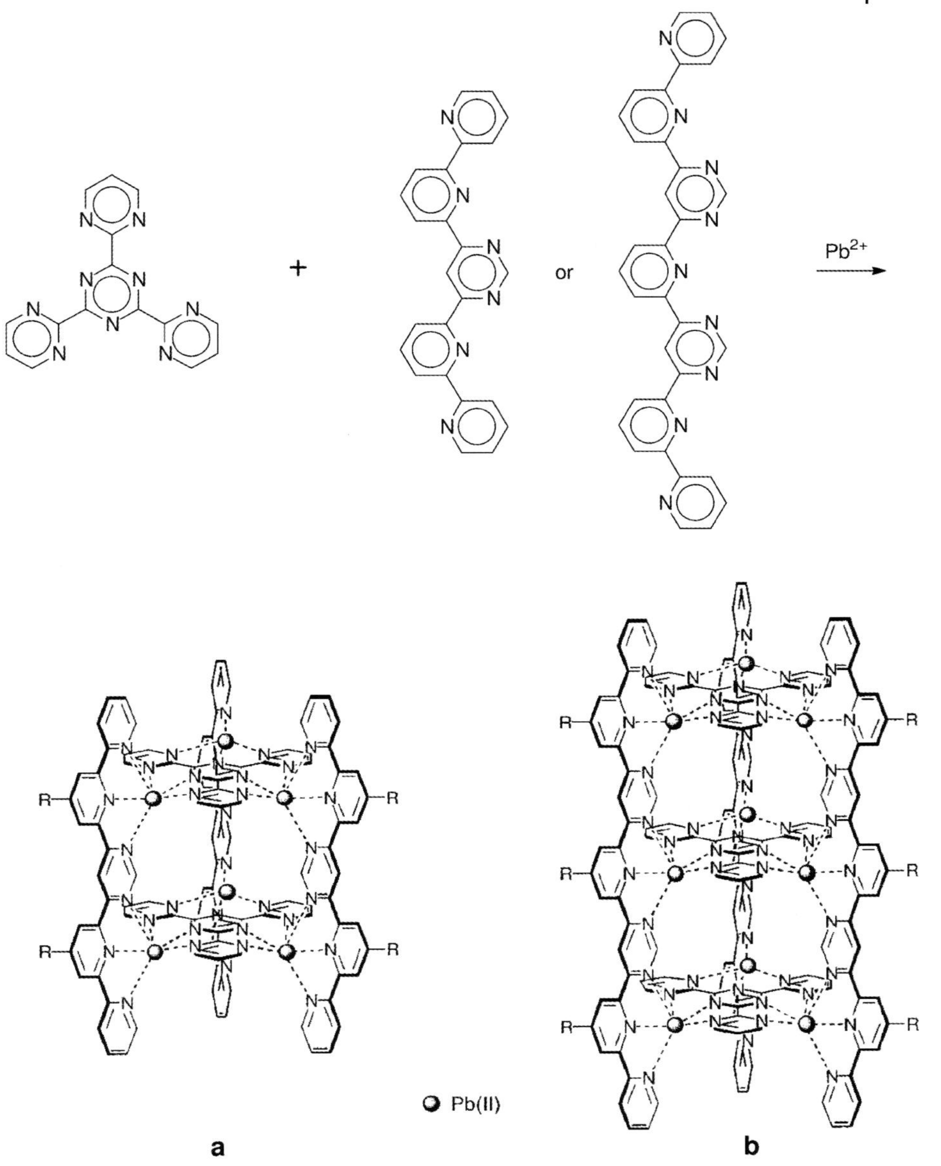

○ Pb(II)

a   b

Scheme 4.11 Cylindrical complex array from the ligands shown.
(Reprinted with permission from [87]).

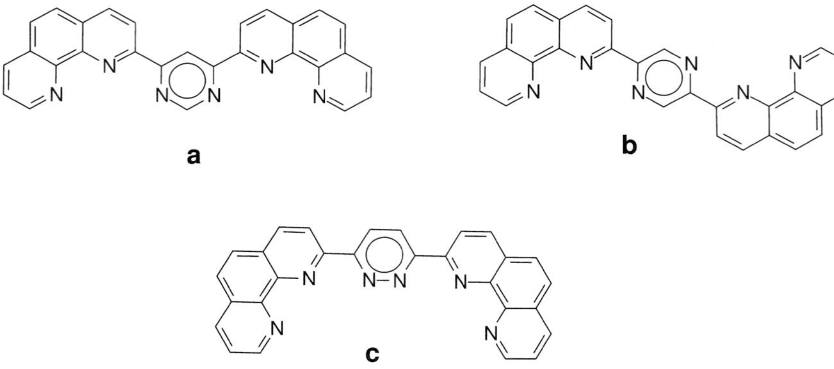

Figure 4.21 Phenanthroline-based *bis*-coordinating ligands [88].

The concept of multiple chelators has been continued by the synthesis of phenanthroline-based systems [88]. *Bis*-tridentate "Lehn-type" ligands as well as molecules in which the chelating moieties are pointing in opposite directions were prepared (Figure 4.21). All the corresponding *mono*- and dimetallic *bis*"terpyridine" Ru(II) complexes could be prepared, except for the pincer-shaped ligand **c**, in which the dinuclear complex was not accessible because of steric hindrance. The crystal structure of the rack-like dinuclear *bis*terpyridine Ru(II) complex was demonstrated to be very similar to the rack structures of *bis*(bipyridinyl)pyridazines [89].

Ligands of the structure shown in Figure 4.21b possessing the central pyrazine ring connected to peripheral phenanthroline moieties were able to form supramolecular assemblies consisting of four ligands and four metal ions. Unlike the "Lehn-type" grids, these structures possess a D_4-symmetry and are therefore chiral [90]. Whereas the unsubstituted ligand (Scheme 4.12a) formed a racemic mixture

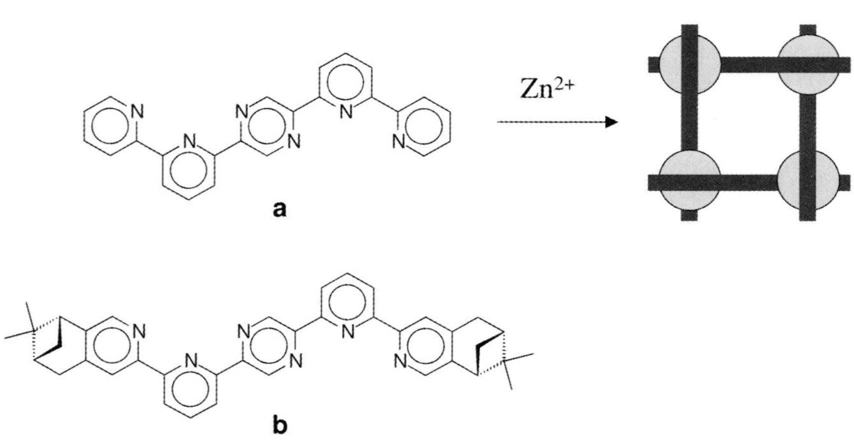

Scheme 4.12 Formation of chiral "grid"-like assemblies [90].

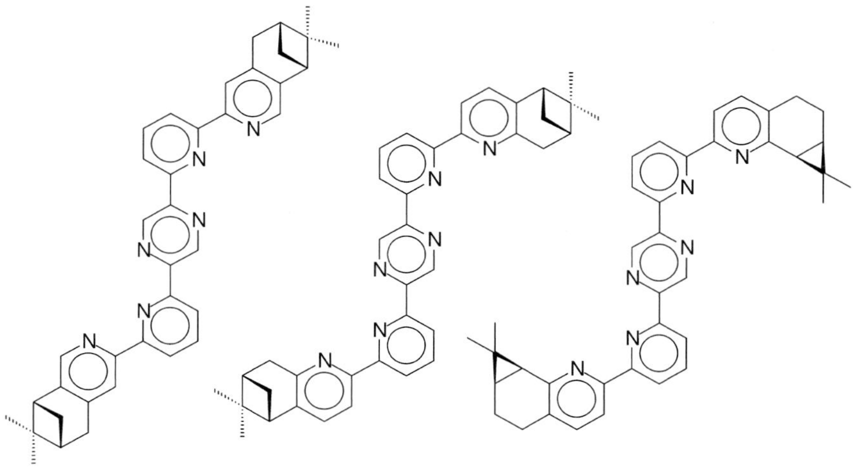

Figure 4.22 *Bis*-coordinating ligand, bearing chiral groups [91].

of supramolecular structures, the self-assembly could be directed in a stereo-selective fashion by attaching chiral pinene groups onto the terminal pyridine rings (Scheme 4.12b). A series of differently substituted ligands was prepared to evaluate the chiral nature of the assembly process (Figure 4.22) [91].

4.3.2
Helicates

A series of oligopyridines was prepared by Abruña et al. (Scheme 4.13b) [92], in which 4′,4′′′′′-*bis*(methylthio)-4′′,4′′′′-*bis*(*n*-propylthio)septipyridine was treated with Cu(II) or Co(II) salts to generate double helices involving two ligand moieties and two metal ions. A mixture of Cu(II)/Cu(I) (1 : 2) led to a trinuclear helix with Cu(II) ions being tetracoordinated (bipyridine motif on the ligand). With Cu(I) ions only, a tetranuclear complex was formed (the final pyridine rings act as monodentate ligands). With an analogous hexaterpyridine, an oxidative switching between a trinuclear and a dinuclear species was shown to be possible (Scheme 4.13a) [93].

In these complexes, the ligand strands act as double terpyridine units, leaving two pyridines uncoordinated. The same research group reported the preparation of 6,6′′-diphenyl-4,4′′-*bis*(alkylthio)-2,2′:6′,2′′-terpyridines, which formed helical complexes with Cu(I) ions [94]. In this case, the terpyridine serves as a *bis*- and *mono*-dentate ligand; with Cu(II) ions, a regular terpyridine complex was formed.

Trinuclear helicates (Figure 4.23) could be obtained by the self-assembly of one *tris*-terpyridine and one *tris*-bipyridine ligand together with three Cu(II) ions through formation of pentacoordinated $[Cu(tpy)(bpy)]^{2+}$ complexes (see Chapter 3) [95].

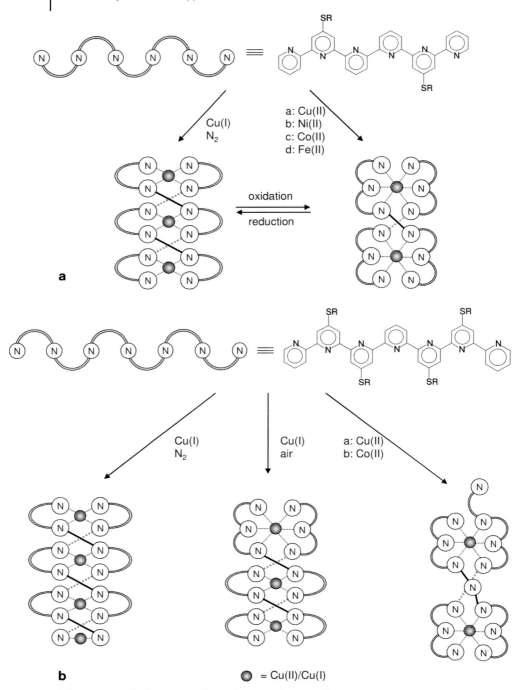

Scheme 4.13 Helical aggregates from oligoterpyridine complexes [93, 94].

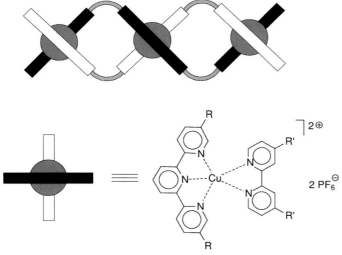

Figure 4.23 Self-assembly to a trinuclear helicate via formation of $[Cu(tpy)(bpy)]^{2+}$ complexes [95].

4.3.3
Rotaxanes and Catenanes

Rotaxanes of 4′-functionalized terpyridine have appeared in a recent publication [96, 97] in which a cationic (dipyridinium)ethane system acts as an axle in a supramolecular system with crown ethers. After the self-assembly process, the system was eventually locked by end-capping the axle with a bulky end-group (Figure 4.24). Unsubstituted 24-crown-8 as well as the dibenzo- and dinaphthyl-analogs were employed. Iron(II) complexes of all ligands that were prepared showed a significant red shift of the metal-ligand charge transfer absorption band for the complexes containing the benzocrown ethers. The crystal structure of the ligand revealed a π-interaction of the benzo-substituent of the crown ether with the terpyridine moiety, causing a stabilization of the MLCT in the complex.

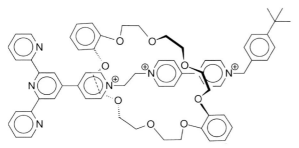

Figure 4.24 Schematic representation of a terpyridine, consisting of a rotaxane assembly with a crown ether [96].

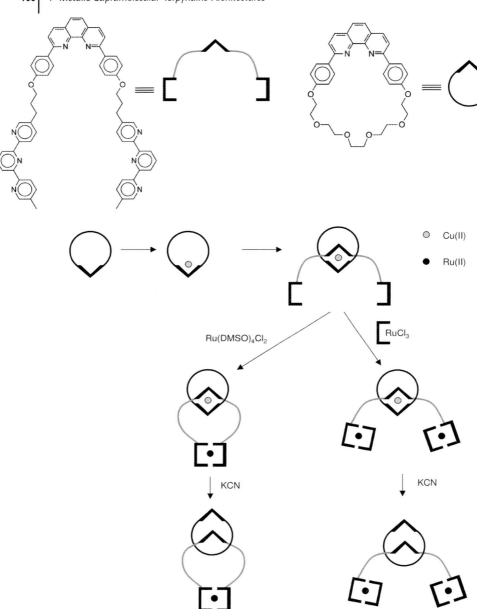

Scheme 4.14 Rotaxanes and catenanes via the self-assembly of tetrahedral phenanthroline followed by the formation of $[Ru(tpy)_2]^{2+}$ complexes [98, 99].

Another example from this research group opened the way to rotaxanes as well as catenanes (Scheme 4.14). Here, a 2,9-*bis*phenylphenanthroline was incorporated into a crown ether ring structure by first generating a Cu(II)-*mono*-terpyridine complex (*bis*-complexes are not possible due to steric hindrance). The addition of the terpyridine-phenanthroline-terpyridine strand can subsequently thread the macrocyclic ring to form a *bis*-phenanthroline Cu(II) complex, which with Ru(DMSO)$_4$Cl$_2$ formed the desired catenane or with [Ru(tpy)]$^{3+}$ fragments to yield the rotaxane. In a final step, the copper ion was removed by addition of KCN to give the free corresponding rotaxane or catenane [98]. The catenane can be remetalated with Zn(II) and Ag(I) ions. Photophysical investigations of all systems revealed electron and energy transfer processes in these compounds with the

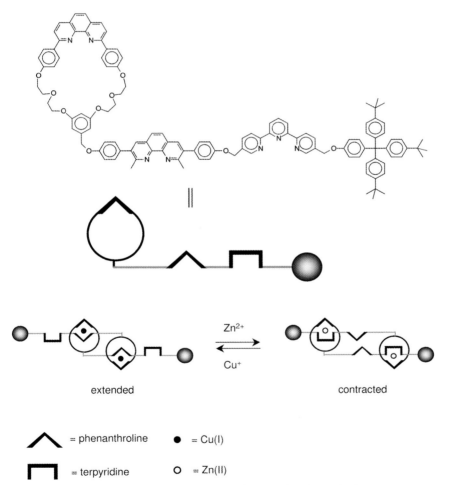

extended   contracted

∧ = phenanthroline   ● = Cu(I)

⊓ = terpyridine   ○ = Zn(II)

Scheme 4.15 "Molecular muscle" by redox-switchable terpyridine-phenanthroline complexes [100, 101].

direction being dependent on the metal ion used. The ligand-centered luminescence was quenched in the Zn(II), Ag(I), and free catenane cases; the opposite effect took place for the Cu(II) catenane. This latter system could be of interest as a potential "molecular switch" [99].

An extended molecule consisting of a phenanthroline moiety within a crown ether-like ring structure, connected to another phenanthroline bound to a terpyridine moiety (bearing a bulky end-group in the 5″-position) in the 5-position, has been prepared and subsequently self-assembled with Cu(I) ions to form a linked *bis*-rotaxane [100]. After exchanging Cu(I) for Zn(II), the coordination switched from a *tetra*coordinated to *penta*coordinated system involving the terpyridine moiety (Scheme 4.15). The resulting reversible extension and contraction represented the first step toward novel applications such as artificial muscles or molecular machines. More information focusing on molecular muscles can be found in [101].

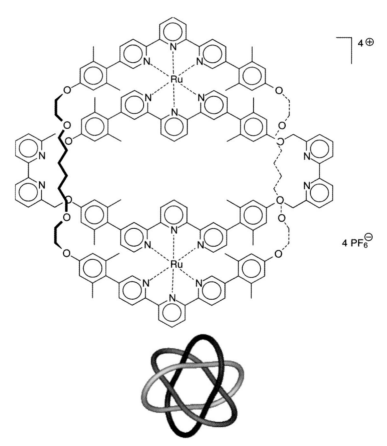

Figure 4.25 Two-ring structure (top) as a first step towards a "Borromean link" (bottom) [102].

Another possible system of interlocked macrocycles is the so-called "Borromean link", represented by three orthogonally interweaved rings, leading to an integral molecular unit without catenation of two rings. The first step towards such systems was performed by connecting two macrocycles via the formation of Ru(II) complexes (Figure 4.25) [102]. The bipyridine ligands on one of the rings were used as the template for the third ring.

4.3.4
Other Assemblies

A novel switchable ruthenium complex is represented by the "scorpionate" system depicted in Scheme 4.16 [103]. A Ru(II) complex was generated from a ligand combination of 1,10-phenanthroline, terpyridine, and benzonitrile. The latter, in turn, is able to coordinate to the Ru(II) metal center, in a "back-biting" manner. The coordinated benzonitrile group can be reversibly replaced by water through either thermal or photochemical induction.

A calix[4]arene possessing one face modified by appending four terpyridine moieties has been prepared to act as a scaffold-like "pre-organizer" for supramolecular architectures (Figure 4.26) [104]. Complexation with Ni(II), Cu(II), or Co(II) led to intramolecular macrocyclization incorporating two metal ions.

A wheel-like, self-assembled structure has been achieved using 6-carboxy-terpyridine, which acts as a tetradentate ligand [105]. With a 2 : 1 ratio of the ligand to europium triflate, a *mono*nuclear *bis*-complex was formed. Addition of more europium ions resulted in self-assembly to the hexanuclear complex of wheel-like structure, where the carboxylates act as bridging ligands between the subunits. An octahedrally coordinated seventh europium ion lies in the center, uniquely coordinating to all six complex units, in essence acting as an axle for this molecular wheel (Scheme 4.17).

Terpyridines, functionalized with pyridine in the 6,6″-positions, led to "metallo-supramolecular zippers" by the addition of metal-ions (Figure 4.27). The metal ion was coordinated to both the terpyridine and pyridine; the two remaining coordination sites are occupied by either MeCN or trifluorosulfonate ions [106].

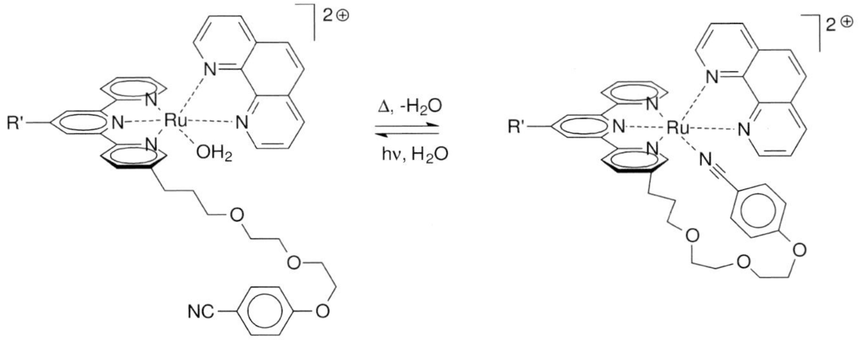

Scheme 4.16 Scorpionate terpyridine complex [103].

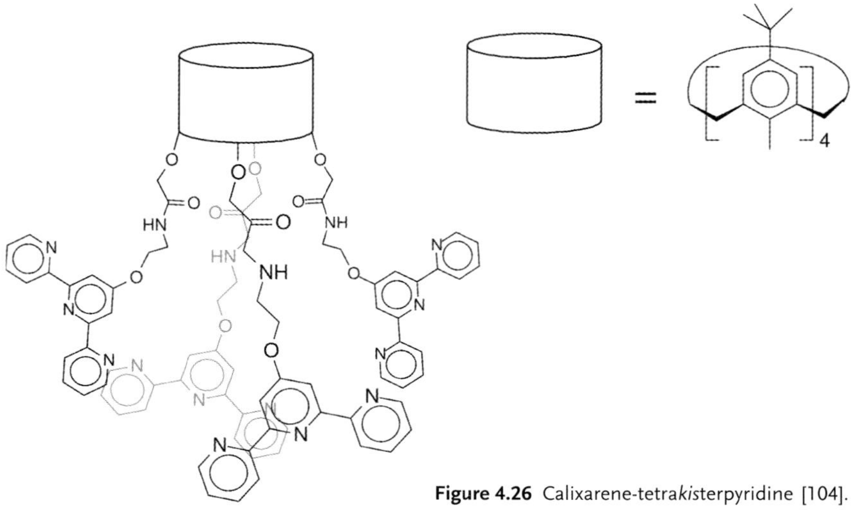

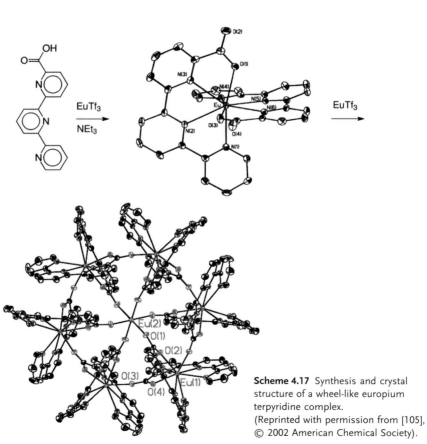

Figure 4.26 Calixarene-tetra*kis*terpyridine [104].

Scheme 4.17 Synthesis and crystal structure of a wheel-like europium terpyridine complex.
(Reprinted with permission from [105], © 2002 American Chemical Society).

4.4
Cycles

Bis-terpyridine ligands with suitable spacers are able to form ring structures by complexation with metal ions. Macrocycles, based on rigid *bis*-terpyridine ligands, were the first reported examples. Whereas linear rigid systems form polymers, ligand systems with an angle in the spacer group result in cyclic complexes. Because of the given geometry, the size of the rings is in many cases predetermined. Synthetically accessible linkers include the rigid ethynyl group, which has been reviewed by Ziessel (Figure 4.28a) [5]. The 60° angle of the two ethynyl-terpyridines [107], connected to the phenanthroline, favor a tricyclic structure; however, the less stable tetracyclic analog was also formed. Recently, Newkome et al. prepared the more compact triangular metallomacrocycle (Figure 4.28c) by a one-step macrocyclization of 1,2-*bis*(terpyridine-4-yl-ethynyl)benzene with either FeCl$_2$ or [Ru(DMSO)$_4$(Cl)$_2$)], or a step-wise process giving mixed metals [108].

In another example, the synthesis of box-like cycles has been reported [109]. The *bis*-terpyridine ligand was obtained by alkylation of the unique N-pyridinyl locus of 4'-(4-pyridyl)terpyridine with 4,4'-*bis*(bromomethyl)biphenyl; subsequent complexation with Fe(II) ions gave rise to a bicyclic compound (Figure 4.28b) as the main product.

Large molecular squares were constructed based on palladium or rhenium complexes in the corners and [Ru(tpy)$_2$]$^{2+}$ complexes as the sides (Scheme 4.18) [110, 111]. 4'-(4-Pyridyl)terpyridine [112] complexes of iron, ruthenium, and osmium were employed in a self-assembly reaction with BrRe(CO)$_5$ or (dppf)Pd(H$_2$O)$_2$(OTf)$_2$ [where dppf = 1,1'-*bis*(diphenylphosphino)ferrocene]. These complexes act as corner points and possess square-planar coordination geometry leading to molecular squares. Multi-electron redox processes were observed in the assemblies, and luminescence was exclusively detected for the osmium-

Figure 4.27 Formation of a "zipper" by complexation of 6,6''-*bis*-pyridino-terpyridine. (Reprinted with permission from [106], © 2004 American Chemical Society).

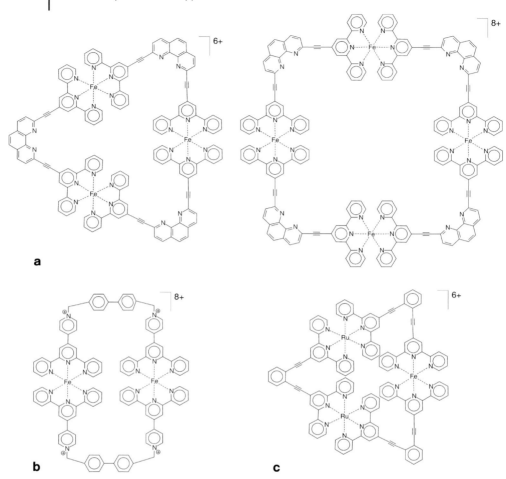

Figure 4.28 Dimeric, trimeric, and tetrameric cycles from rigid *bis*-terpyridines [5, 108, 109].

containing square. Hydrogen-bonding and π-stacking interactions associated with 4′-(4-pyridinyl)-2,2′:6′,2″-terpyridine have led to interesting zipper-like supramolecular structures, for example with Cd(II) [113].

An interesting macrocycle was presented by Newkome et al. [114–116]. A meta-*bis*(terpyridinyl)benzene [117] ring was complexed with Fe(II) and Ru(II or III), leading to a hexameric cycle. This geometry is favored because of the 120° angle present in the ligand. Two approaches were used to obtain the ruthenium cycle (Scheme 4.19). In a self-assembly process in which stoichiometric amounts of free ligand were reacted with a di-Ru(III) complex of the same or different ligand (route a), the hexamer was obtained. In the second approach (route b), the macrocycle was constructed by a step-wise sequence via a "hemisphere" as the intermediate.

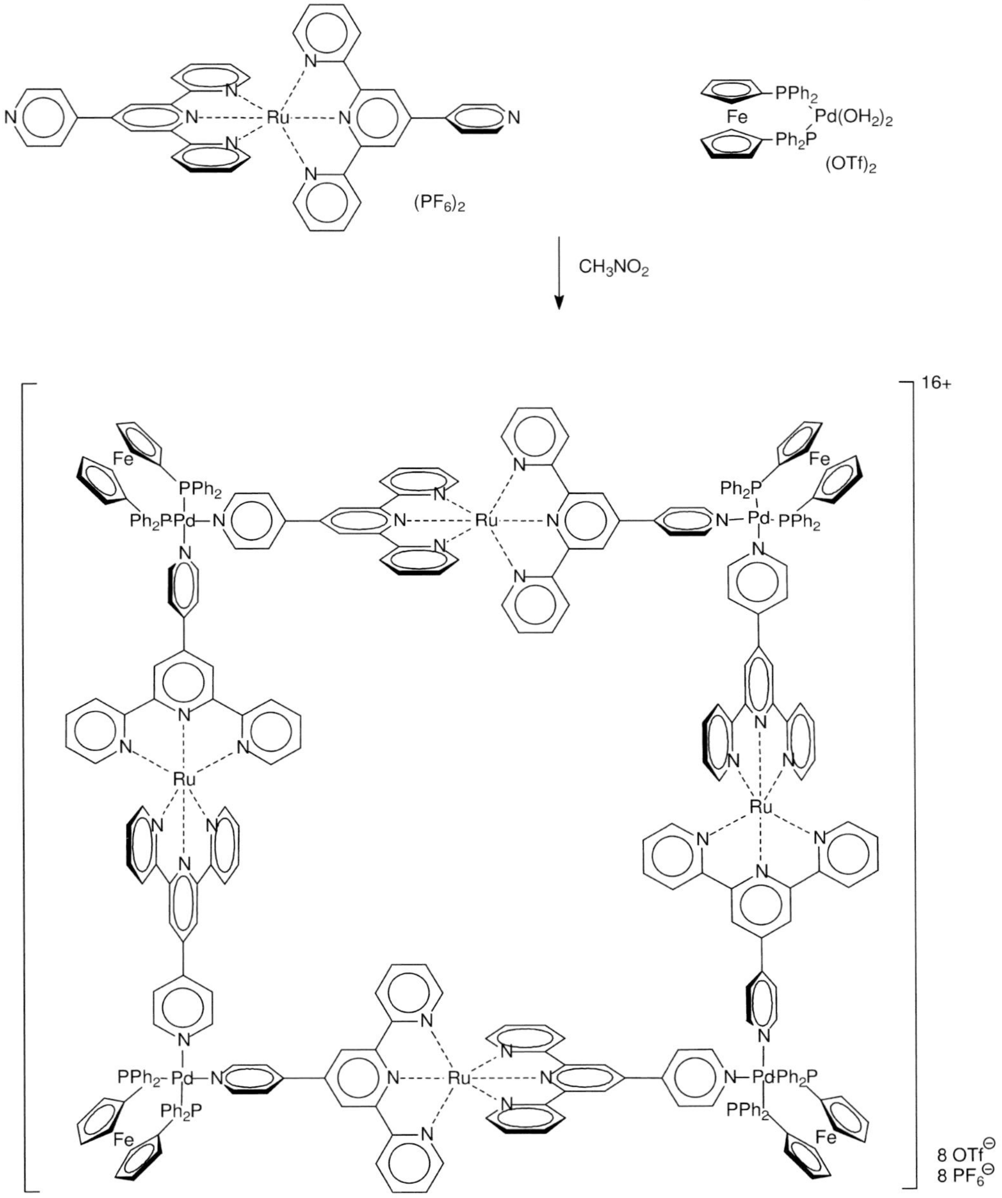

Scheme 4.18 Multimetallic supramolecular squares [110, 111].

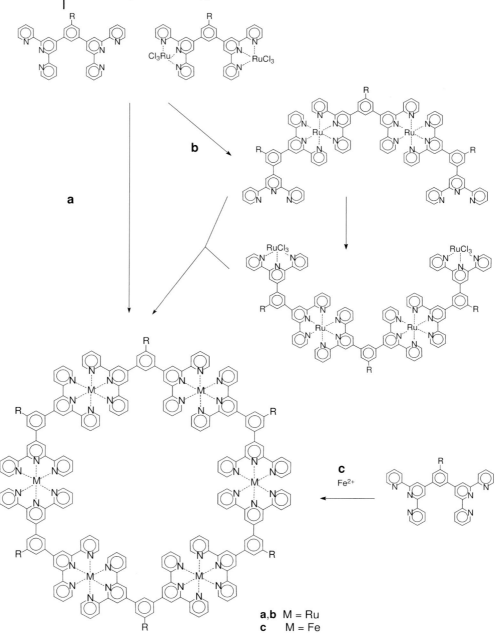

Scheme 4.19 Hexameric macrocycle by Newkome et al. [114].

Moreover, an analogous cycle was prepared by simply adding Fe(II) ions or [Ru(DMSO)$_4$(Cl)$_2$] to the ligand, leading to exclusive formation of the hexameric macrocycle. As well as standard techniques and cyclic voltammetry, TEM was also used to prove the existence of the macrocycles. The utilization of this assembly process has led to a new approach to nanofabrication, as shown in Scheme 4.20 [118].

So far, most macrocycles have been constructed from building blocks possessing a rigid spacer, in which the direction of complexation is also predetermined by the angle within the spacer group; therefore, macrocyclic assemblies were obtained exclusively in well-defined sizes.

The formation of macrocyclic products from *bis*-terpyridine ligands linked by flexible spacers is much more challenging, because these telechelics allow the formation of different-sized rings as well as polymers. Despite these diffi-culties, different flexible groups have been employed in the synthesis of rings. Because of the inherent flexibility, a mixture of various rings and coordination polymers were obtained in most cases. Therefore, low concentrations had to be applied in order to push the ring-chain equilibrium to favor ring formation. Furthermore, purification of the product by column chromatography mixture was necessary.

Mononuclear macrocycles can only be formed if the spacer possesses (a) suffi-cient flexibility and (b) appropriate length. In the following example (Figure 4.29a) from Moore et al. [119], the terpyridines are linked in the 5-position: 1,3-*bis*(2,2′:6′,2″-terpyridinyl-5-ylmethylsulfanyl)propane and 1,4-*bis*(2,2′:6′,2″-ter-pyridinyl-5-ylmethylsulfanyl)butane. The spacer length was predicted to be optimal by molecular modeling; thus, upon the addition of Fe(II) or Ni(II) ions, an intramolecular cyclization was realized and verified by X-ray structure analysis. When adding the free ligand to a solution of the Fe(II) complex of unfunctionalized terpyridine, ligand exchange was observed, resulting in the formation of macro-cyclic complexes. This behavior can be explained by the chelate effect. In a similar manner, three terpyridine moieties were connected to a cyclic triamine (Figure 4.29b) [120]; complexation with Eu(III) led to a room-temperature luminescent bicyclic complex involving all terpyridine groups. In all of these cases, the terpyridines were connected in the 5-position. Because of this, the connecting points of the resulting complex are in close proximity; therefore, even short spacers allow a successful complexation.

4′-Functionalized terpyridines, where the functionalities are on the opposite faces of the octahedral complex, require much longer spacers or a special geometry. The system (Figure 4.29c) in which the terpyridine moieties are linked via tri(ethylene glycol) spacers to a rigid, central 5,5′-*bis*(phenyl)bipyridine has been synthesized by Constable et al. [121]. Flexible chains protrude at 120° angles from the rod-like central unit; thus, the terpyridine groups are able to complex the metal ions, allowing the formation of macrocycles (as confirmed by X-ray crystal structure analysis). The bipyridine unit is accessible for further complexation in order to construct extended supramolecular systems. In continuing experiments, a macrocycle containing a 2,7-(diethylene glycol)-2,7-naphthalene linker, has been

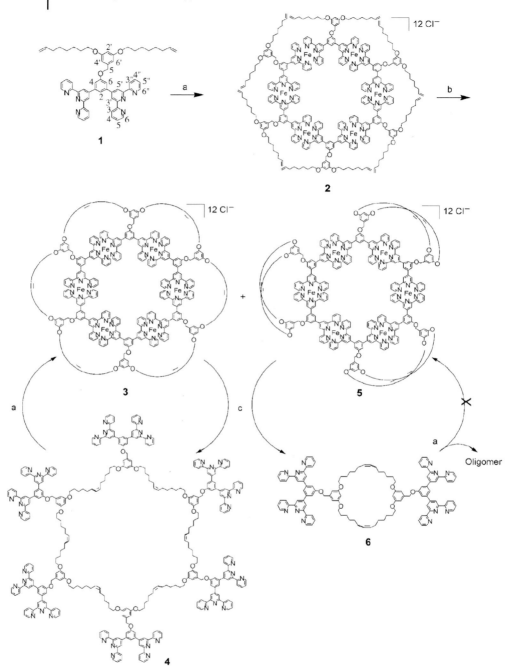

Scheme 4.20 Nanofabrication – assembly-disassembly of iron hexameric macrocycles. (Reprinted with permission from [118]).

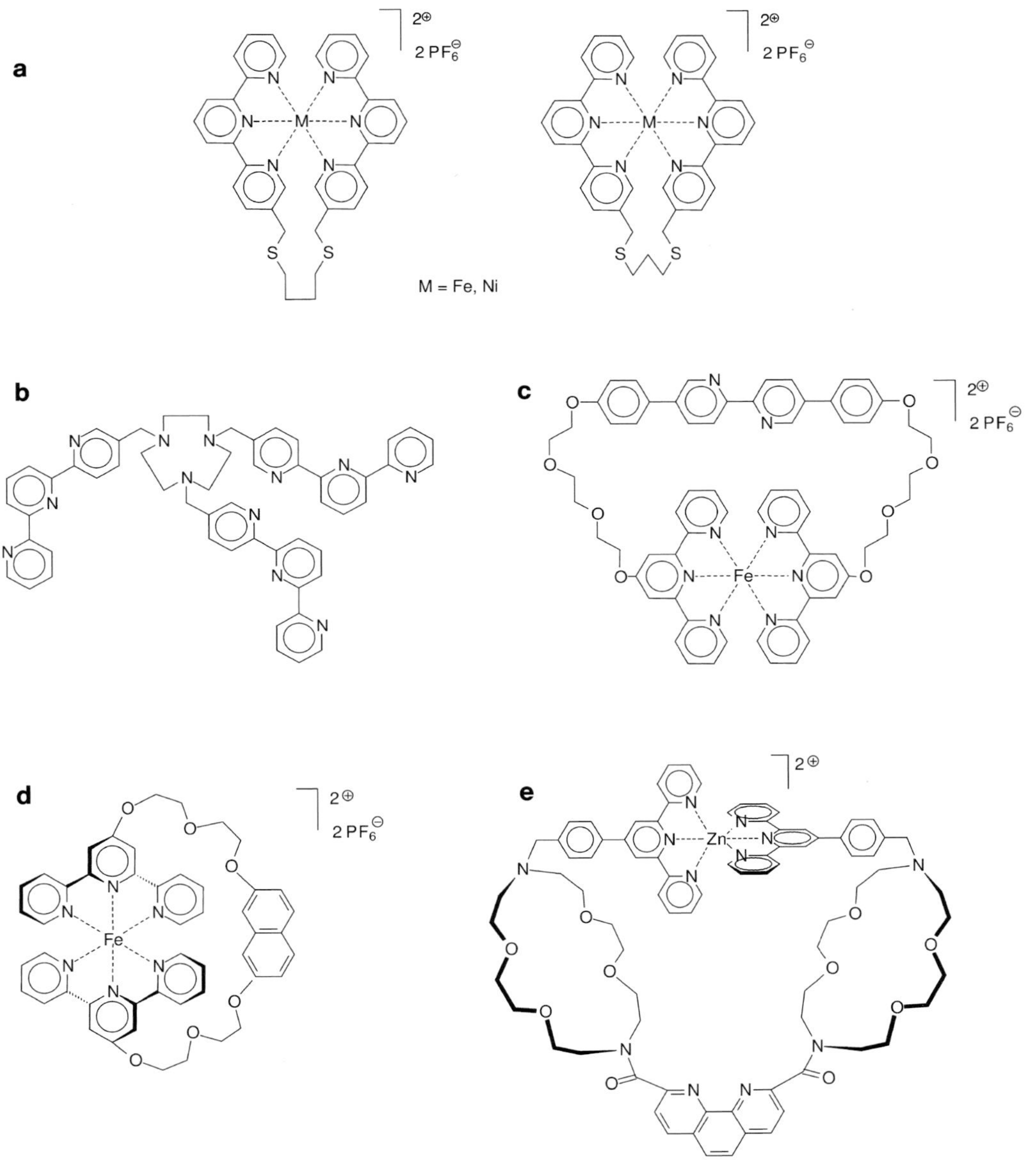

Figure 4.29 Mononuclear macrocycles of 5- and 4′-linked *bis*-terpyridines [119–123].

prepared (Figure 4.29d); the "handle" is conformationally locked in the cleft between the terpyridine moieties, resulting in a chiral structure [122]. This hairpin helicate was further expanded by the use of 2,2′:6′,2″:6″,2‴:6‴,2⁗-quinque-pyridine [124] termini to generate double-helicates and trifold knots [125].

Another mononuclear ring structure was obtained from a multifunctional ligand (Figure 4.29e). An extended polytopic ligand containing two terpyridine moieties that were linked via azacrown ether moieties to a central phenanthroline unit was complexed with Zn(II) ions to form the mononuclear complex [123]. In a different experiment, the terpyridine end groups were complexed to $[Ru(bpy)_2]^{2+}$ fragments.

A recent report by Constable et al. described the synthesis of a family of macrocycles [126]. The terpyridines were linked by flexible triethylene glycol chains in the 4'-position; since the linker is too short for macrocyclization incorporating a single metal center, only tri- and tetranuclear rings were obtained accompanied by some polymeric material (Figure 4.30). To minimize the amount of polymeric product, the reaction was conducted under dilute conditions in which a low (0.4 M) concentration was maintained; the products were isolated by column chromatography. The composition of these cyclic products was analyzed by ESI-MS; furthermore, a shift of the ^{1}H NMR signals of the terpyridine protons, when compared to the corresponding polymer, was found and attributed to the different chemical environment of the aromatic protons. Similar macrocycles, in this case with hexyl-spacers linking the terpyridine moieties, were recently prepared [127]. The tri- and tetranuclear macrocycles were isolated and identified by MALDI-TOF MS. A *bis*terpyridine with a hexadecyloxy spacer was also subjected to Fe(II) ions, from which aggregates of up to [9 + 9], probably macrocycles, were detected by MALDI-TOF MS, but the [2 + 2] macrocycle was shown to be the main product [128]. Constable et al. further elucidated the formation of homo-$[RuL_2]^{4+}$ and heteroleptic $[RuLL']^{4+}$ metallomacrocycles, where L represents different PEGs possessing capped terpyridine moieties [129].

The first examples of metallodendritic spiranes (spirometallodendrimers) were obtained in the group of Newkome [130] via incorporation of one terpyridine unit within each dendritic quadrant of dendrimers derived from the pentaerythritol core (Figure 4.31). These moieties were subsequently complexed with either Fe(II) or Ru(II) ions, exclusively leading to intramolecular cyclization of the spiro type.

In a recent publication, a terpyridine within a pentaamine macrocycle (connected at the 6,6''-positions) has been constructed (Figure 4.32) [131]. Because of the geometry, with the metal-coordinating site pointing inside the ring, traditional $[M(tpy)_2]^{2+}$ complexes cannot be formed; however, hexa-coordinated metal complexes are generated based on two ligands and two metal ions with an OH$^-$ ion as a bridging ligand, leading to a dimeric structure, in which π–π stacking interactions are present. Treatment of this *bis*-complex with acid showed that the secondary amine groups were protonated first, followed by protonation of the terpyridine nitrogens with strong acid.

In an analogous system, a crown ether ring was attached to terpyridine by means of chiral pinene groups (Figure 4.32, bottom); amino acid derivatives were shown to enantioselectively bind, as guests, to this macrocycle. The resulting fluorescence quenching makes this macrocycle a useful sensor for host-guest binding [132]. Other ethereally bridged terpyridines have been reported, such as a 5,5''-bridged terpyridine with a 34-membered ring encompassing a "*bis*phenol A" moiety [133].

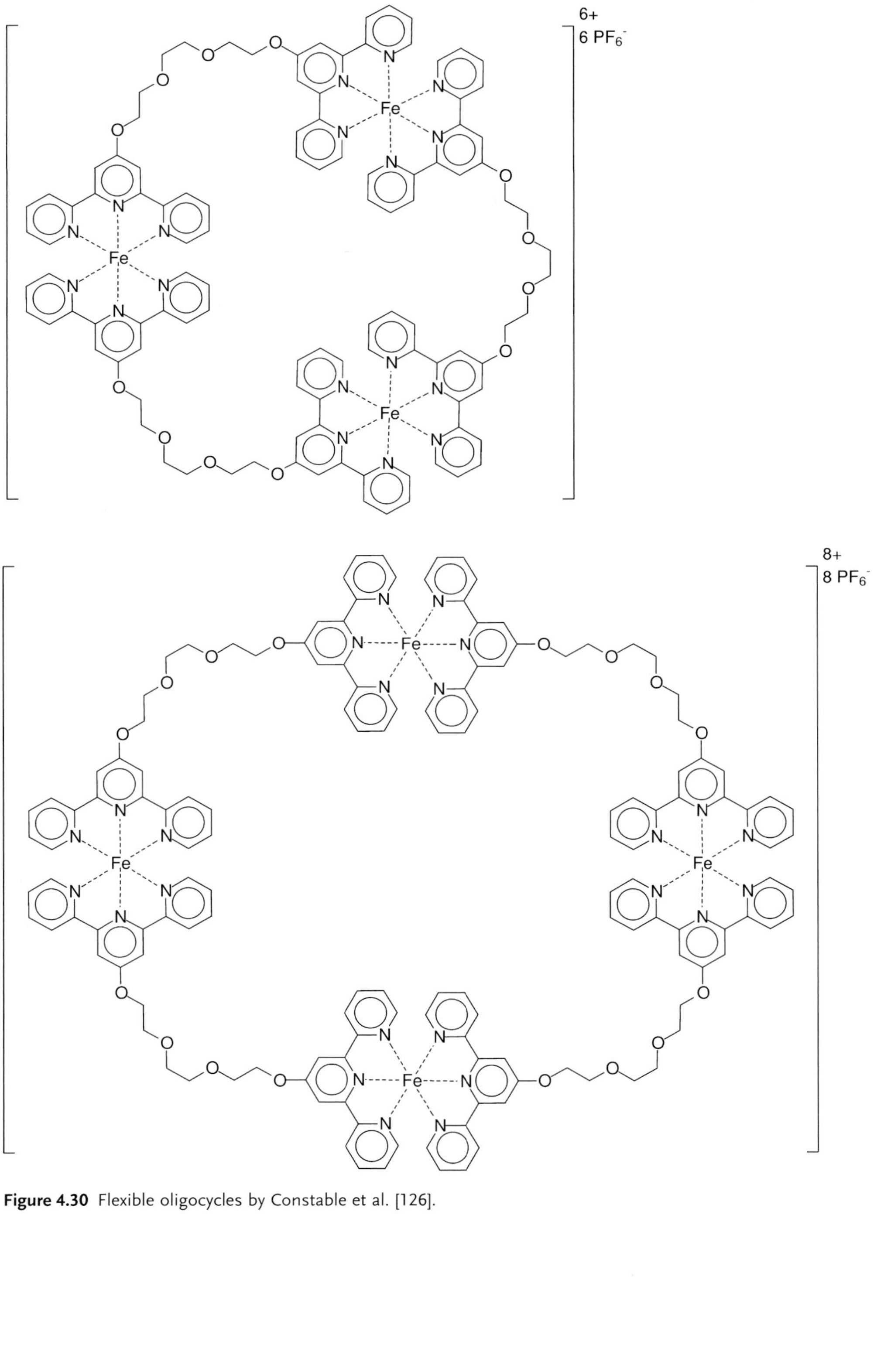

Figure 4.30 Flexible oligocycles by Constable et al. [126].

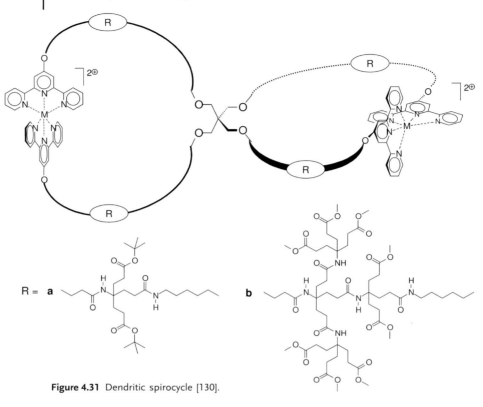

R = **a**   **b**

Figure 4.31 Dendritic spirocycle [130].

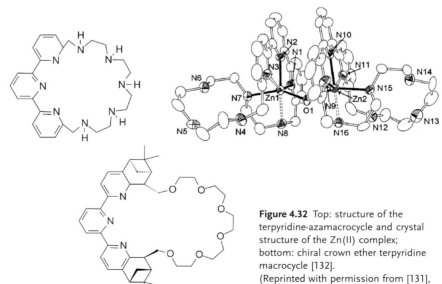

Figure 4.32 Top: structure of the terpyridine-azamacrocycle and crystal structure of the Zn(II) complex; bottom: chiral crown ether terpyridine macrocycle [132].
(Reprinted with permission from [131], © 2004 American Chemical Society).

Fullerene Terpyridine Complexes

Fullerenes are of special importance in contemporary chemistry because of their photophysical and electrochemical properties [134, 135], thus leading to the development of novel molecular electronics or light-harvesting devices. Most of the current research into these particular kinds of potential solar cells include fullerenes because of their electron-accepting properties. Therefore, the combination of fullerenes with $[Ru(tpy)_2]^{2+}$ complexes has a special significance. The pioneering work in this field was performed by Constable et al. and Diederich et al. [136, 137]. Terpyridines, bearing a malonate group linked by *oligo*(ethylene glycol)s, were prepared in two steps (Figure 4.33) utilizing a Bingel reaction of brominated malonates with C_{60} (loss of the bromine and addition to an unsaturated fullerene bond) leading to the fullereno-terpyridines. In an alternative approach, the terpyridine was directly attached to the fullerene by adding a brominated ethyl terpyridinylacetate under Bingel conditions to C_{60}. The functional ligands were subsequently converted to the corresponding Ru(II) complexes by traditional Ru(III)/(II) chemistry. In the case where the units are spatially close, a strong interaction between the fullerene and the terpyridine complex was found by cyclic voltammetry.

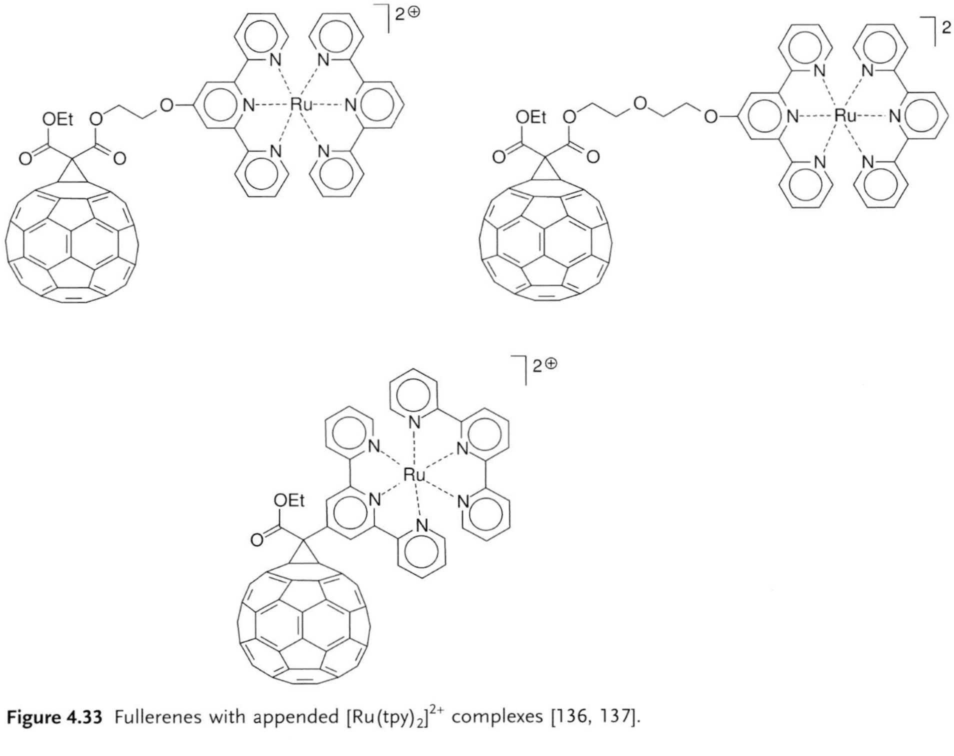

Figure 4.33 Fullerenes with appended $[Ru(tpy)_2]^{2+}$ complexes [136, 137].

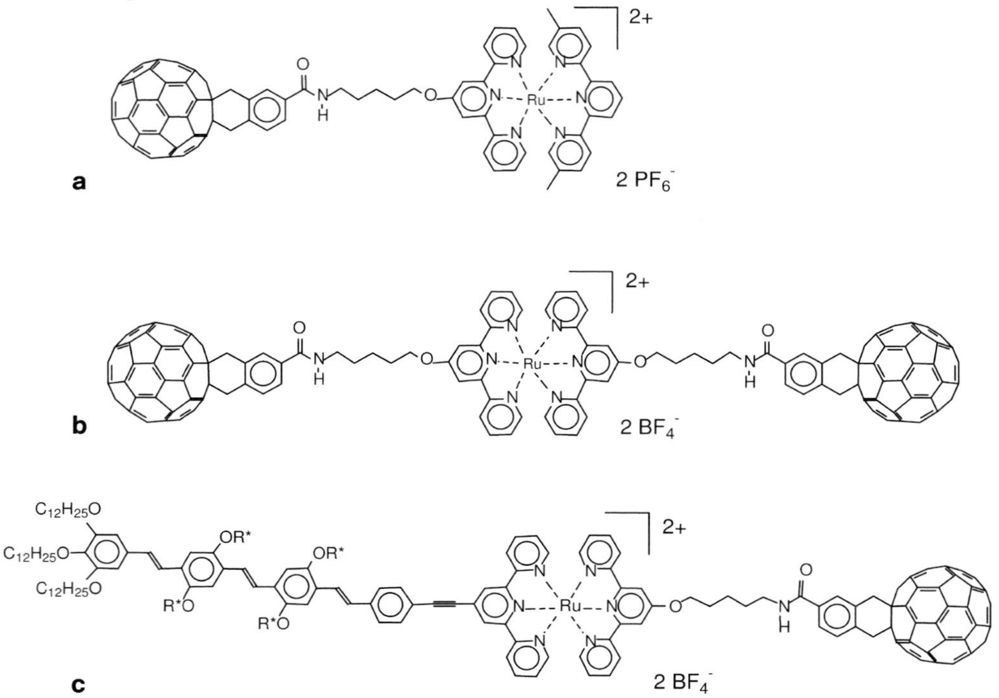

Figure 4.34 Fullerene-[Ru(tpy)$_2$]$^{2+}$ dyads and triads [138, 139].

The approach was continued by Schubert et al. [138]: a functionalized fullerene (obtained in several steps, which included a Diels-Alder reaction) containing an acid chloride was reacted with an aminoterpyridine, resulting in a terpyridinyl-fullerene linked by an alkyl group. AB-dyads and ABA-triads were prepared by complexing the ligand with an [Ru(tpy)(X)$_3$] complex or a second equivalent of the functional ligand and RuCl$_3$ (Figure 4.34a and b).

Another molecular moiety that has been studied intensively in the context of organic solar cells is the highly luminescent and electron-donating *oligo(p-phenylene vinylene)* (OPV) that has already been covalently linked to fullerenes [140]. Recently, the combination of these two units has been achieved via the formation of a [Ru(tpy)$_2$]$^{2+}$ complex leading to an ABC-triad (Figure 4.34c) [139]. Photoinduced absorption experiments revealed a lifetime of the charge-separated state of less than 100 μs (this is the time scale of the experiment – the actual lifetime could not be determined). Energy transfer from the OPV to the fullerene resulted in the quenching of the OPV luminescence.

Li et al. demonstrated the conversion of 4′-(*p*-tolyl)terpyridine to the corresponding α-azidomethyl derivative [141]. The cycloaddition of this azide to the fullerene resulted in a ring-opening of the fullerene cage [142, 143] (1,6-aza-bridged, Scheme 4.21). Subsequent complexation with [Ru(ttpy)$_2$(Cl)$_3$], where ttpy = 4′-(*p*-tolyl)terpyridine, led to the fullerene-[Ru(ttpy$_2$]$^{2+}$ complex dyad.

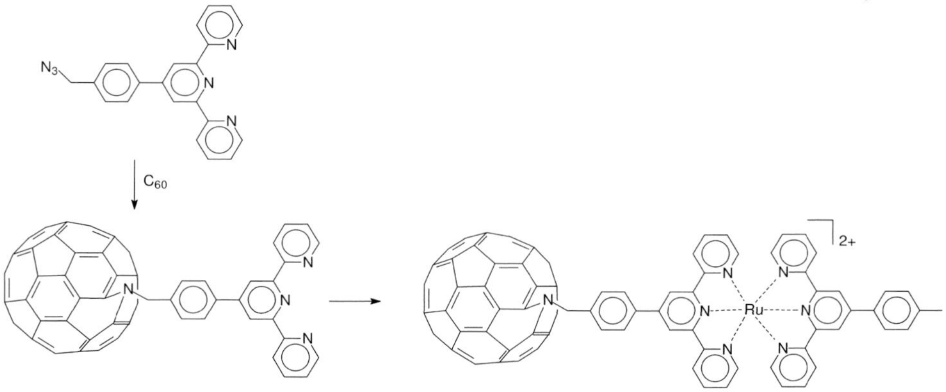

Scheme 4.21 Coupling of an azide-functionalized terpyridine to a fullerene [141].

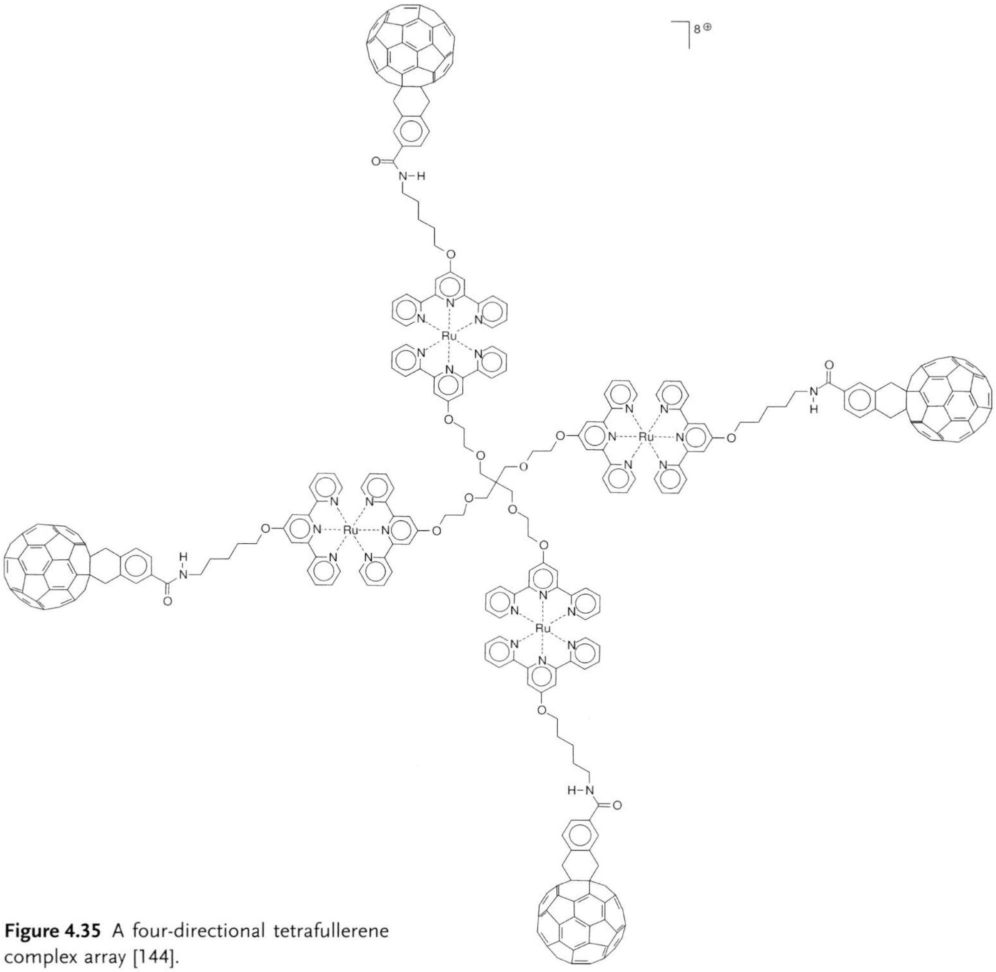

Figure 4.35 A four-directional tetrafullerene complex array [144].

A *tetrafullerene* adduct was successfully prepared by the treatment of *tetra-terpyridine* core, based on pentaerythritol, with the terpyridine-containing fullerene derivative (Figure 4.35) to give the four-directional complex with buckyballs in the periphery [144].

4.6
Complexes Containing Biochemical Groups

An interesting approach to complexes containing biofunctionality is a combination of terpyridine complex chromophores with biomolecules, which can subsequently act as luminescent labels for biological processes. Furthermore, the capability of electron transfer could potentially be of help in the of study electron transfer processes in biological systems.

Ziessel et al. reported the synthesis of a terpyridine containing an L-tyrosine group (Figure 4.36) and the subsequent formation of the corresponding Ru(II) complexes [145]. This combination was chosen for future studies of "Photosystem II" (in which water is oxidized to oxygen), a large membrane-bound protein system in the photosynthetic reaction center of green plants. It could also help in the construction of artificial water oxidation catalysts. An ethynyl group was chosen as the linker, together with the Pd-catalyzed coupling of the iodide of the tyrosine group with the 4′-triflate-terpyridine [146]. As well as the compound containing a free phenolic group, a complex also bearing a protected group was prepared. While a solution of the unprotected complex in MeCN showed luminescence even at ambient temperature, it was quenched in a K_2CO_3-containing DMF solution, indicating an electron transfer from the phenolic group of the tyrosine moiety by photoexcitation in basic conditions, followed by back-electron-transfer. In the corresponding protected compound, in which the phenol is blocked, no electron transfer is possible, and, as a result, emission was observed.

Cyclodextrin "cups" (monosaccharides attached to terpyridines as well as the corresponding complexes have been described in Chapter 2, Table 2.1, e.g., **1p** and **1q**) were employed to build supramolecular systems consisting of a 4′-tolyl-terpyridine ruthenium complex and a guest-binding moiety (Scheme 4.22) [147].

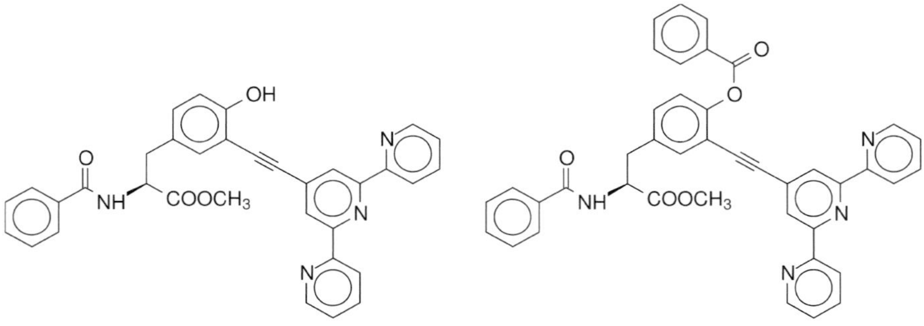

Figure 4.36 Terpyridines bearing an L-tyrosine group [145].

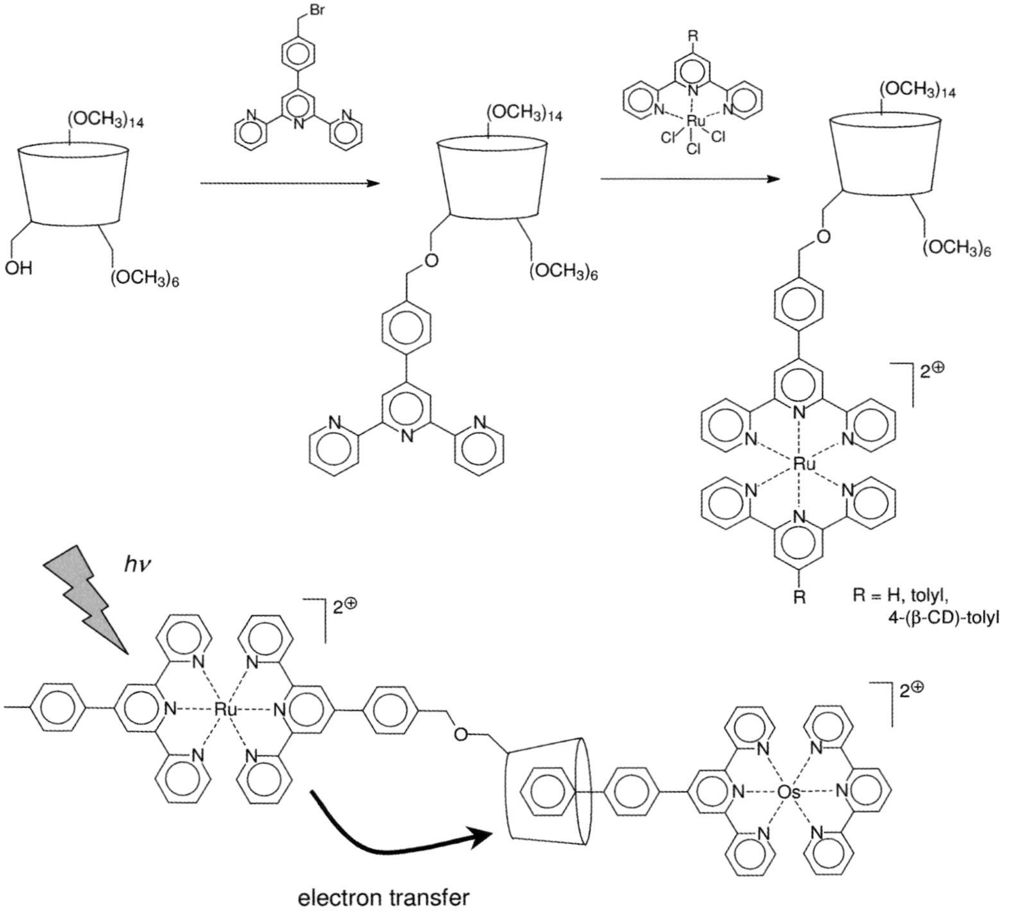

Scheme 4.22 Cyclodextrin-functionalized [Ru(ttpy)$_2$]$^{2+}$ complexes binding with a [Os(4′-biphenyl-tpy)(tpy)]$^{2+}$ guest, leading to an electron- transfer [147, 148].

All but one of the hydroxymethyl groups of the cyclodextrin were protected before nucleophilic reaction with 4′-[p-(bromomethyl)phenyl]terpyridine in a Williamson-type ether coupling. This functionalized ligand was subsequently complexed with either [Ru(tpy)(Cl)$_3$] or [Ru(ttpy)(Cl)$_3$], respectively, the latter being known to be a room-temperature emitter. The final product may act as an optical sensor for the hydrophobic binding of guests to the cyclodextrin, as shown by the addition of anthraquinone-2-carboxylic acid. A quenching of the phosphorescence was observed, resulting from an intermolecular electron transfer from the Ru(II) complex to the quinone.

In further studies [also including a *bis*(cyclodextrin) complex], different quinone guests were used leading to a luminescence quenching of different intensities depending on the quinone [148]. In another experiment, an [Os(biphenyl-tpy)(tpy)]$^{2+}$ complex (biphenyl group as the guest) was added to the host system. In the resulting Ru(II)-Os(II) dyad, an electron transfer from Ru(II) to Os(II) could be observed after oxidation of the osmium center. An Ru(II)-Os(II) energy transfer was, however, not detected because of the short lifetime of the ruthenium emission (in accordance with covalent dyads linked by saturated spacers). This energy transfer could be achieved by the use of long-lifetime complexes [149]. In the present case, a [Ru(bpy)$_3$]$^{2+}$ complex bearing cyclodextrins was synthesized (Figure 4.37).

The binding of terpyridine metal complexes containing a guest-binding unit (biphenyl or adamantyl) resulted in an energy transfer. Moreover, the direction of the energy transfer could be reversed, depending on the metal center of the guest-

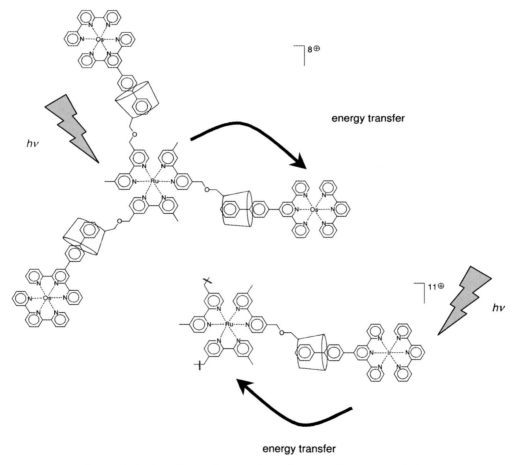

Figure 4.37 Binding of terpyridine complexes of Os(II) and Ir(III) to a cyclodextrin-functionalized [Ru(bpy)$_3$]$^{+2}$ complex enabling energy- transfer processes [149].

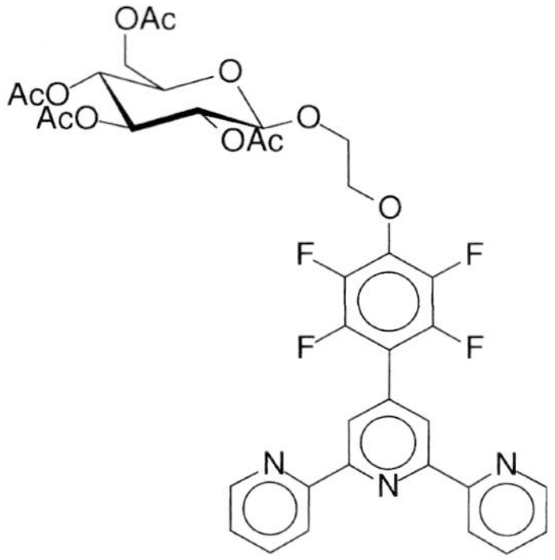

Figure 4.38 Glucose-functionalized terpyridine ligand [150].

complex; whereas Os(II) complexes lead to an energy-transfer from Ru(II) to Os(II) (core to periphery), the direction is reversed by Ir(III) complexes, as guests. The energy transfer from the three peripheral complexes to the Ru(II) core makes such systems interesting potential light-harvesting devices.

The novelty of this host-guest system, compared to the ones previously described (dyads), is that both donor and acceptor are being brought together spatially via non-covalent interactions.

Besides cyclodextrin, the monosaccharide, glucose, was also attached to a terpyridine ligand (Figure 4.38). This sugar unit was introduced via reaction of a 1-bromoethyl-functionalized glucose with a 4'-(p-hydroxytetrafluorophenyl)ter-pyridine [150].

Various porphyrin $[Ru(tpy)_2]^{2+}$ complex conjugates, AB dyads, ABA, BAB, and ABC triads have been constructed and investigated by Flamigni et al., in which the Ru(II) center was shown to be a strong electron and energy acceptor, quenching the porphyrin fluorescence [20, 151–154]. The porphyrin site has been studied in both the free-base and the metalated form with either Zn(II) or Au(II) ions (Figure 4.39). Zinc porphyrins with one or two $[Ru(tpy)_2]^{2+}$ and $[Os(tpy)_2]^{2+}$ complexes appended have been synthesized and their nonlinear optical properties studied [155].

A photoinduced two-step electron- and energy-transfer could be observed. The singlet excited state porphyrin is first quenched by the ruthenium component, leading to an Ru(II) complex localized triplet excited state (^{3}MLCT) which, in turn, transfers its triplet energy back to the porphyrin unit attached to it so as to generate the porphyrin triplet excited state. These authors also utilized terpyridine Ir(III) complexes within the porphyrin triad system, showing a multi-step electron transfer from the free or Zn(II) complexed porphyrin via the Ir(III) complex to the Au(III) complexed porphyrin [156, 157].

In a recent contribution, an $[Ru(tpy)_2]^{2+}$ porphyrin dyad has been reported in which the two parts were linked by a p-phenylene group, and a phenylethynyl moiety was attached to the other side of the terpyridine ruthenium complex; such a combination was shown to prolong the triplet lifetime of the ruthenium complex.

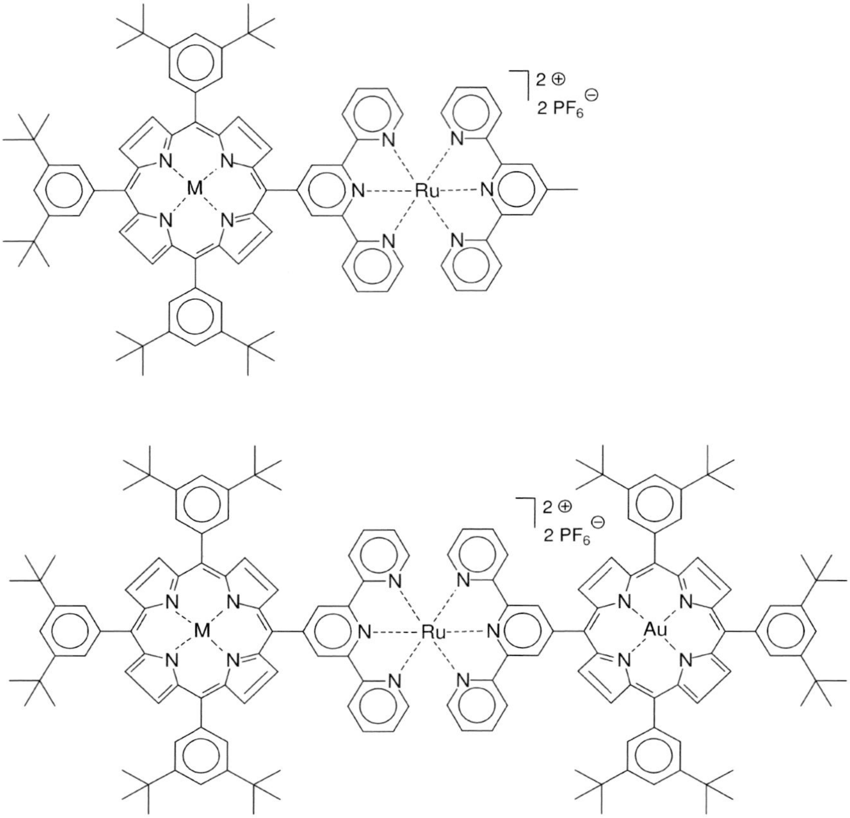

a M = Zn
b M = 2 H

Figure 4.39 Porphyrin – [M(tpy)$_2$]$^{2+}$ complexes [151–154].

This compound permitted a more detailed study of the energy-transfer processes in the dyad [158].

Another approach in combining biochemistry with terpyridine supramolecular chemistry is demonstrated by the coupling of biotin to 4′-aminoterpyridine by means of the well-known isocyanate coupling reaction [159, 160]. As well as a short alkyl spacer, a polymeric poly(ethylene glycol) chain has also been introduced (Figure 4.40). Biotin is known to bind strongly to the protein avidin via multiple *H*-bonding with a geometry comparable to a "lock and key" system.

A dinuclear iron carbonyl complex was attached to an [Ru(tpy)$_2$]$^{2+}$ complex, mimicking the active site of iron-hydrogenase (Figure 4.41) [161, 162]. Irradiation of this complex resulted in an electron-transfer, which could be utilized to produce hydrogen.

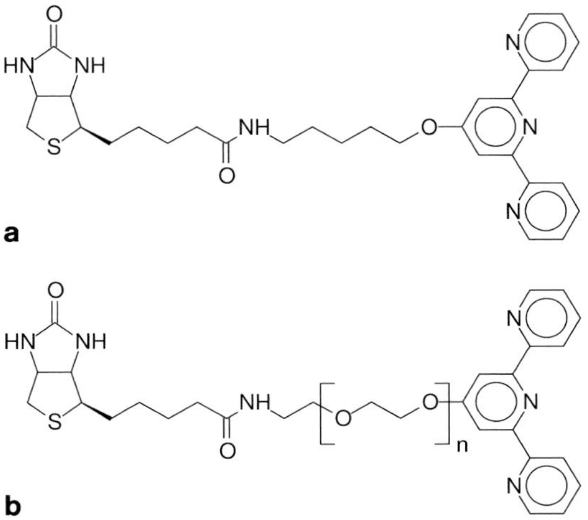

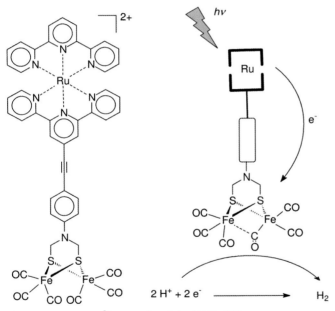

a

b

Figure 4.40 Biotin-functionalized terpyridine ligands [159, 160].

Figure 4.41 [Ru(tpy)$_2$]$^{2+}$-iron carbonyl dyad [161, 162].

A connection of DNA fragments to terpyridine complexes has been reported in which two complementary DNA sequences (20 base pairs) were connected via a tri(ethylene glycol) to a terpyridine moiety, respectively, and subsequently the corresponding symmetric Ru(II) complexes were prepared [163]. The mixing of these two complexes resulted in long linear arrays through self-assembly, in which the length could be adjusted by the molar ratios of the ligands within the mixture.

References

1 J. P. Sauvage, J. P. Collin, J. C. Chambron, S. Guillerez, C. Coudret, V. Balzani, F. Barigelletti, L. De Cola, L. Flamigni, *Chem. Rev.* **1994**, *94*, 993–1019.

2 E. Baranoff, J.-P. Collin, L. Flamigni, J.-P. Sauvage, *Chem. Soc. Rev.* **2004**, *33*, 147–155.

3 E. A. Medlycot, G. S. Hanan, *Chem. Soc. Rev.* **2005**, *34*, 133–142.

4 F. Barigelletti, L. Flamigni, *Chem. Soc. Rev.* **2000**, *29*, 1–12.

5 R. Ziessel, *Synthesis* **1999**, 1839–1865.

6 A. Harriman, A. Khatyr, R. Ziessel, A. C. Benniston, *Angew. Chem. Int. Ed.* **2000**, *39*, 4287–4290.

7 L. Hammarstroem, F. Barigelletti, L. Flamigni, M. T. Indelli, N. Armaroli, G. Calogero, M. Guardigli, A. Sour, J.-P. Collin, J.-P. Sauvage, *J. Phys. Chem. A* **1997**, *101*, 9061–9069.

8 A. Harriman, R. Ziessel, *Chem. Commun.* **1996**, 1707–1716.

9 V. Grosshenny, A. Harriman, R. Ziessel, *Angew. Chem. Int. Ed.* **1995**, *34*, 1100–1102.

10 V. Grosshenny, A. Harriman, R. Ziessel, *Angew. Chem. Int. Ed.* **1995**, *34*, 2705–2708.

11 V. Grosshenny, F. M. Romero, R. Ziessel, *Inorg. Chem.* **1997**, *62*, 1491–1500.

12 M. Hissler, A. El-Ghayoury, A. Harriman, R. Ziessel, *Angew. Chem. Int. Ed.* **1998**, *37*, 1717–1720.

13 A. Khatyr, R. Ziessel, *J. Org. Chem.* **2000**, *65*, 3126–3134.

14 A. C. Benniston, V. Grosshenny, A. Harriman, R. Ziessel, *Angew. Chem. Int. Ed.* **1994**, *33*, 1884–1885.

15 A. Harriman, M. Hissler, A. Khatyr, R. Ziessel, *Eur. J. Inorg. Chem.* **2003**, 955–959.

16 A. C. Benniston, V. Grosshenny, A. Harriman, R. Ziessel, *Dalton Trans.* **2004**, 1227–1232.

17 F. Barigelletti, L. Flamigni, V. Balzani, J.-P. Collin, J.-P. Sauvage, A. Sour, *New J. Chem.* **1995**, *19*, 793–798.

18 F. Barigelletti, L. Flamigni, G. Calogero, L. Hammarstrom, J.-P. Sauvage, J.-P. Collin, *Chem. Commun.* **1998**, 2333–2334.

19 J.-P. Collin, P. Lainé, J.-P. Launay, J.-P. Sauvage, A. Sour, *Chem. Commun.* **1993**, 434–435.

20 J.-P. Collin, A. Harriman, V. Heitz, F. Odobel, J.-P. Sauvage, *J. Am. Chem. Soc.* **1994**, *116*, 5679–5690.

21 J.-P. Collin, P. Gaviña, V. Heitz, J.-P. Sauvage, *Eur. J. Org. Chem.* **1998**, 1–14.

22 F. Barigelletti, L. Flamigni, V. Balzani, J.-P. Collin, J.-P. Sauvage, A. Sour, E. C. Constable, A. M. W. C. Thompson, *J. Am. Chem. Soc.* **1994**, *116*, 7692–7699.

23 M. T. Indelli, F. Scandola, J.-P. Collin, J.-P. Sauvage, A. Sour, *Inorg. Chem.* **1996**, *35*, 303–312.

24 H. Torieda, A. Yoshimura, K. Nozaki, S. Sakai, T. Ohno, *J. Phys. Chem. A* **2002**, *106*, 11034–11044.

25 A. El-Ghayoury, A. Harriman, A. Khatyr, R. Ziessel, *J. Phys. Chem. A* **2000**, *104*, 1512–1523.

26 A. Harriman, A. Khatyr, R. Ziessel, *Dalton Trans.* **2003**, 2061–2068.

27 A. C. Benniston, S. Mitchell, S. A. Rostron, S. Yang, *Tetrahedron Lett.* **2004**, *45*, 7883–7885.

28 A. C. Benniston, A. Harriman, P. Li, C. A. Sams, M. D. Ward, *J. Am. Chem. Soc.* **2004**, *126*, 13630–13631.

29 A. C. Benniston, A. Harriman, P. Li, C. A. Sams, *Tetrahedron Lett.* **2003**, *44*, 4167–4169.

30 A. C. Benniston, A. Harriman, P. Li, C. A. Sams, *J. Am. Chem. Soc.* **2005**, *127*, 2553–2564.

31 T.-Y. Dong, M.-C. Lin, M. Y.-N. Chiang, J.-Y. Wu, *Organometallics* **2004**, *23*, 3921–3930.

32 T.-Y. Dong, H.-W. Shih, L.-S. Chang, *Langmuir* **2004**, *20*, 9340–9347.

33 H.-W. Shih, T.-Y. Dong, *Inorg. Chem. Commun.* **2004**, *7*, 646–649.

34 R. Ziessel, V. Grosshenny, M. Hissler, C. Stroh, *Inorg. Chem.* **2004**, *43*, 4262–4271.

35 A. C. Benniston, G. M. Chapman, A. Harriman, S. A. Rostron, *Inorg. Chem.* **2005**, *44*, 4029–4036.

36 R. Ziessel, C. Stroh, *Tetrahedron Lett.* **2004**, *45*, 4051–4055.

37 G. D. Storrier, S. B. Colbran, *Inorg. Chim. Acta* **1999**, *284*, 76–84.

38 G. D. Storrier, S. B. Colbran, D. C. Craig, *J. Chem. Soc., Dalton Trans.* **1997**, 3011–3028.

39 A. Amini, K. Bates, A. C. Benniston, D. J. Lawrie, E. Soubeyrand-Lenoir, *Tetrahedron Lett.* **2003**, *44*, 8245–8247.

40 S. Baitalik, X. Wang, R. H. Schmehl, *J. Am. Chem. Soc.* **2004**, *126*, 16304–16305.

41 B. Hasenknopf, J. Hall, J.-M. Lehn, V. Balzani, A. Credi, S. Campagna, *New. J. Chem.* **1996**, *20*, 725–730.

42 G. Lowe, A. S. Droz, T. Vilaivan, G. W. Weaver, J. J. Park, J. M. Pratt, L. Tweedale, L. R. Kelland, *J. Med. Chem.* **1999**, *42*, 3167–3174.

43 N. J. Wheate, J. G. Collins, *Coord. Chem. Rev.* **2003**, *241*, 133–145.

44 M. J. Clarke, *Coord. Chem. Rev.* **2002**, *232*, 69–93.

45 K. v. d. Schilden, F. Garcìa, H. Kooijman, A. L. Spek, J. G. Haasnoot, J. Reedijk, *Angew. Chem. Int. Ed.* **2004**, *43*, 5668–5670.

46 J. A. Barron, S. Glazier, S. Bernhard, K. Takada, P. L. Houston, H. D. Abruña, *Inorg. Chem.* **2003**, *42*, 1448–1455.

47 S. Bernhard, K. Takada, D. J. Diaz, H. D. Abruña, H. Mürner, *J. Am. Chem. Soc.* **2001**, *123*, 10265–10271.

48 S. Bernhard, J. I. Goldsmith, K. Takada, H. D. Abruña, *Inorg. Chem.* **2003**, *42*, 4389–4393.

49 F. Barigelletti, L. Flamigni, M. Guardigli, J.-P. Sauvage, J.-P. Collin, A. Sour, *Chem. Commun.* **1996**, 1329–1330.

50 F. Loiseau, R. Passalacqua, S. Campagna, M. I. J. Polson, Y.-Q. Fang, G. S. Hanan, *Photochem. Photobiol. Sci.* **2002**, *1*, 982–990.

51 B. Galland, D. Limosin, H. Laguitton-Pasquier, A. Deronzier, *Inorg. Chem. Commun.* **2002**, *5*, 5–8.

52 T. Akasaka, J. Otsuki, K. Araki, *Chem. Eur. J.* **2002**, *8*, 130–136.

53 T. Akasaka, T. Mutai, J. Otsuki, K. Araki, *Dalton Trans.* **2003**, 1537–1544.

54 E. C. Constable, C. E. Housecroft, E. Schofield, S. Encinas, N. Armaroli, F. Barigelletti, L. Flamigni, E. Figgemeier, J. G. Vos, *Chem. Commun.* **1999**, 869–870.

55 E. C. Constable, R. W. Handel, C. E. Housecroft, A. F. Morales, L. Flamigni, F. Barigelletti, *Dalton Trans.* **2003**, 1220–1222.

56 S. Encinas, L. Flamigni, F. Barigelletti, E. C. Constable, C. E. Housecroft, E. R. Schofield, E. Figgemeier, D. Fenske, M. Neuburger, J. G. Vos, M. Zehnder, *Chem. Eur. J.* **2002**, *8*, 137–150.

57 A. Harriman, A. Mayeux, A. De Nicola, R. Ziessel, *Phys. Chem. Chem. Phys.* **2002**, *4*, 2229–2235.

58 C. Ringenbach, A. De Nicola, R. Ziessel, *J. Org. Chem.* **2003**, *68*, 4708–4719.

59 A. De Nicola, C. Ringenbach, R. Ziessel, *Tetrahedron Lett.* **2003**, *44*, 183–187.

60 A. Barbieri, B. Ventura, F. Barigelletti, A. De Nicola, M. Quesada, R. Ziessel, *Inorg. Chem.* **2004**, *43*, 7359–7368.

61 T. Akasaka, H. Inoue, M. Kuwabara, T. Mutai, J. Otsuki, K. Araki, *Dalton Trans.* **2003**, 815–821.

62 R. Argazzi, E. Bertolasi, C. Chiorboli, C. A. Bignozzi, M. K. Itokazu, N. Y. M. Iha, *Inorg. Chem.* **2001**, *40*, 6885–6891.

63 J.-D. Lee, L. M. Vrana, E. R. Bullock, K. J. Brewer, *Inorg. Chem.* **1998**, *37*, 3575–3580.

64 G. S. Hanan, D. Volkmer, U. S. Schubert, J.-M. Lehn, G. Baum, D. Fenske, *Angew. Chem. Int. Ed.* **1997**, *36*, 1842–1844.

65 G. S. Hanan, U. S. Schubert, D. Volkmer, E. Rivière, J. M. Lehn, N. Kyritsakas, J. Fischer, *Can. J. Chem.* **1997**, *75*, 169–182.

66 G. S. Hanan, C. R. Arana, J. M. Lehn, D. Fenske, *Angew. Chem. Int. Ed.* **1995**, *34*, 1122–1124.

67 M. Benaglia, F. Ponzini, C. R. Woods, J. S. Siegel, *Org. Lett.* **2001**, *3*, 967–969.

68 G. S. Hanan, C. R. Arana, J. M. Lehn, G. Baum, D. Fenske, *Chem. Eur. J.* **1996**, *2*, 1292–1302.

69 A. Credi, V. Balzani, S. Campagna, G. S. Hanan, C. R. Arana, J.-M. Lehn, *Chem. Phys. Lett.* **1995**, *243*, 102–107.

70 P. Ceroni, A. Credi, V. Balzani, S. Campagna, G. S. Hanan, C. R. Arana, J.-M. Lehn, *Eur. J. Inorg. Chem.* **1999**, 1409–1414.

71 M.-T. Youinou, N. Rahmouny, J. Fischer, J. A. Osborn, *Angew. Chem. Int. Ed.* **1992**, *31*, 733.

72 P. N. W. Baxter, J. M. Lehn, J. Fischer, M.-T. Youinou, *Angew. Chem. Int. Ed.* **1994**, *33*, 2284–2286.

73 U. S. Schubert, C. Eschbaumer, *Org. Lett.* **1999**, *1*, 1027–1029.

74 U. S. Schubert, C. Eschbaumer, *J. Incl. Phenom. Macrocycl. Chem.* **1999**, *35*, 101–109.

75 O. Waldmann, J. Hassmann, P. Muller, G. S. Hanan, D. Volkmer, U. S. Schubert, J. M. Lehn, *Phys. Rev. Lett.* **1997**, *78*, 3390–3393.

76 E. Breuning, M. Ruben, J.-M. Lehn, F. Renz, Y. Garcia, V. Ksenofontov, P. Gütlich, E. Wegelius, K. Rissanen, *Angew. Chem. Int. Ed.* **2000**, *39*, 2504–2507.

77 E. Breuning, G. S. Hanan, F. J. Romero-Salguero, A. M. Garcia, P. N. W. Baxter, J.-M. Lehn, E. Wegelius, K. Rissanen, H. Nierengarten, A. van Dorsselaer, *Chem. Eur. J.* **2002**, *8*, 3458–3466.

78 A. M. Garcia, F. J. Romero-Salguero, D. M. Bassani, J.-M. Lehn, G. Baum, D. Fenske, *Chem. Eur. J.* **1999**, *5*, 1803–1808.

79 M. Barboiu, G. Vaughan, R. Graff, J.-M. Lehn, *J. Am. Chem. Soc.* **2003**, *125*, 10257–10265.

80 H. Nierengarten, E. Leize, E. Breuning, A. Garcia, F. Romero-Salguero, J. Rojo, J.-M. Lehn, A. van Dorsselaer, *J. Mass. Spectrom.* **2002**, *37*, 56–62.

81 D. M. Bassani, J.-M. Lehn, K. Fromm, D. Fenske, *Angew. Chem. Int. Ed.* **1998**, *37*, 2364–2367.

82 D. M. Bassani, J.-M. Lehn, S. Serroni, F. Puntoriero, S. Campagna, *Chem. Eur. J.* **2003**, *9*, 5936–5946.

83 L. Kovbasyuk, H. Pritzkow, R. Krämer, *Eur. J. Inorg. Chem.* **2005**, 894–900.

84 E. Breuning, U. Ziener, J.-M. Lehn, E. Wegelius, K. Rissanen, *Eur. J. Inorg. Chem.* **2001**, 1515–1521.

85 M. Ruben, U. Ziener, J.-M. Lehn, V. Ksenofontov, P. Gütlich, G. B. M. Vaughan, *Chem. Eur. J.* **2005**, *11*, 84–100.

86 U. Ziener, J.-M. Lehn, A. Mourran, M. Möller, *Chem. Eur. J.* **2002**, *8*, 951–957.

87 A. M. Garcia, D. M. Bassani, J.-M. Lehn, G. Baum, D. Fenske, *Chem. Eur. J.* **1999**, *5*, 1234–1238.

88 D. Brown, S. Muranjan, Y. Jang, R. Thummel, *Org. Lett.* **2002**, *4*, 1253–1256.

89 D. Brown, R. Zong, R. P. Thummel, *Eur. J. Org. Chem.* **2004**, 3269–3272.

90 T. Bark, M. Düggeli, H. Stoeckli-Evans, A. von Zelewsky, *Angew. Chem. Int. Ed.* **2001**, *40*, 2848–2851.

91 T. Bark, H. Stoeckli-Evans, A. von Zelewsky, *J. Chem. Soc., Perkin Trans. 1* **2002**, 1881–1886.

92 K. T. Potts, M. Keshavarz-K, F. S. Tham, K. A. G. Raiford, C. Arana, H. D. Abruña, *Inorg. Chem.* **1993**, *32*, 5477–5484.

93 K. T. Potts, M. Keshavarz-K, F. S. Tham, H. D. Abruña, C. Arana, *Inorg. Chem.* **1993**, *32*, 4436–4439.

94 K. T. Potts, M. Keshavarz-K, F. S. Tham, H. D. Abruña, C. Arana, *Inorg. Chem.* **1993**, *32*, 4450–4456.

95 B. Hasenknopf, J.-M. Lehn, G. Baum, D. Fenske, *Proc. Natl. Acad. Sci. USA* **1996**, *93*, 1397–1400.

96 G. J. E. Davidson, S. J. Loeb, *Dalton Trans.* **2003**, 4319–4323.

97 J.-P. Sauvage, *Chem. Commun.* **2005**, 1507–1510.

98 D. J. Cárdenas, P. Gaviña, J. P. Sauvage, *J. Am. Chem. Soc.* **1997**, *119*, 2656–2664.

99 D. J. Cárdenas, J.-P. Colli, P. Gaviña, J. P. Sauvage, D. C. A., J. Fischer, N. Armaroli, L. Flamigni, V. Vicinelli, V. Balzani, *J. Am. Chem. Soc.* **1999**, *121*, 5481–5488.

100 M. C. Jimenez-Molero, C. Dietrich-Buchecker, J.-P. Sauvage, *Chem. Eur. J.* **2002**, *8*, 1456–1466.

101 M. C. Jimenez-Molero, C. Dietrich-Buchecker, J. P. Sauvage, *Chem. Commun.* **2003**, 1613–1616.

102 J. C. Loren, M. Yoshizawa, R. F. Haldimann, L. A., J. S. Siegel, *Angew. Chem. Int. Ed.* **2003**, *42*, 5702–5705.

103 E. R. Schofield, J.-P. Collin, N. Gruber, J.-P. Sauvage, *Chem. Commun.* **2003**, 188–189.

104 Y. Molard, H. Parrot-Lopez, *Tetrahedron Lett.* **2002**, *43*, 6355–6358.

105 Y. Bretonniere, M. Mazzanti, J. Pecaut, M. M. Olmstead, *J. Am. Chem. Soc.* **2002**, *124*, 9012–9013.

106 M. Barboiu, E. Petit, G. Vaughan, *Chem. Eur. J.* **2004**, *10*, 2263–3370.

107 F. M. Romero, R. Ziessel, A. Dupont-Gervais, A. Van Dorsselaer, *Chem. Commun.* **1996**, 551–553.

108 S.-H. Hwang, C. N. Moorefield, F. R. Fronczek, O. Lukoyanova, L. Echegoyen, G. R. Newkome, *Chem. Commun.* **2005**, 713–715.

109 E. C. Constable, E. Schofield, *Chem. Commun.* **1998**, 403–404.

110 S.-S. Sun, A. J. Lees, *Inorg. Chem.* **2001**, *40*, 3154–3160.

111 S.-S. Sun, A. S. Silva, I. M. Brinn, A. J. Lees, *Inorg. Chem.* **2000**, *39*, 1344–1345.

112 E. C. Constable, A. M. W. C. Thompson, *J. Chem. Soc., Dalton Trans.* **1992**, 2947.

113 J. Granifo, M. T. Garland, R. Baggio, *Inorg. Chem. Commun.* **2004**, *7*, 77–81.

114 G. R. Newkome, T. J. Cho, C. N. Moorefield, R. Cush, P. S. Russo, L. A. Godínez, M. J. Saunders, P. Mohapatra, *Chem. Eur. J.* **2002**, *8*, 2946–2954.

115 G. R. Newkome, T. J. Cho, C. N. Moorefield, G. R. Baker, M. J. Saunders, R. Cush, P. S. Russo, *Angew. Chem. Int. Ed.* **1999**, *38*, 3717–3721.

116 G. R. Newkome, T. J. Cho, C. N. Moorefield, P. P. Mohapatra, L. A. Godínez, *Chem. Eur. J.* **2004**, *10*, 1493–1500.

117 P. Wang, C. N. Moorefield, G. R. Newkome, *Org. Lett.* **2004**, *6*, 1197–1200.

118 P. Wang, C. N. Moorefield, G. R. Newkome, *Angew. Chem. Int. Ed.* **2004**, *44*, 1679–1683.

119 P. M. Gleb U. Priimov, P. K. Maritim, P. K. Butalanyi, N. W. Alcock, *J. Chem. Soc., Dalton Trans.* **2000**, 445–449.

120 N. W. Alcock, A. J. Clarke, W. Errington, A. M. Josceanu, P. Moore, S. C. Rawle, P. Sheldon, S. M. Smith, M. L. Turonek, *Supramol. Chem.* **1996**, *6*, 281–291.

121 C. B. Smith, E. C. Constable, C. E. Housecroft, B. M. Kariuki, *Chem. Commun.* **2002**, 2068–2069.

122 H. S. Chow, E. C. Constable, C. E. Housecroft, M. Neuburger, *J. Chem. Soc., Dalton Trans.* **2003**, 4568–4569.

123 F. Loiseau, C. D. Pietro, S. Serroni, S. Campagna, A. Licciardello, A. Manfredi, G. Pozzi, S. Quici, *Inorg. Chem.* **2001**, *40*, 6901–6909.

124 E. C. Constable, I. A. Hougen, C. E. Housecroft, M. Neuberger, S. Schaffner, L. A. Whall, *Inorg. Chem. Commun.* **2004**, *7*, 1128–1131.

125 E. C. Constable, E. Figgemeier, I. A. Hougen, C. E. Housecroft, M. Neuburger, S. Schaffner, L. A. Whall, *Dalton Trans.* **2005**, 1168–1175.

126 E. C. Constable, C. E. Housecroft, C. B. Smith, *Inorg. Chem. Commun.* **2003**, *6*, 1011–1013.

127 P. R. Andres, U. S. Schubert, *Synthesis* **2004**, 1229–1238.

128 P. R. Andres, U. S. Schubert, *Macromol. Rapid Commun.* **2004**, *25*, 1371–1375.

129 E. C. Constable, C. E. Housecroft, M. Neuburger, S. Schaffner, C. B. Smith, *Dalton Trans.* **2005**, 2259–2267.

130 G. R. Newkome, K. S. Yoo, C. N. Moorefield, *Chem. Commun.* **2002**, 2164–2165.

131 C. Bazzicalupi, A. Bencini, A. Berni, A. Bianchi, A. Danesi, C. Giorgi, B. Valtancoli, C. Lodeiro, J. C. Lima, F. Pina, M. A. Bernardo, *Inorg. Chem.* **2004**, *43*, 5134–5146.

132 W.-L. Wong, K.-H. Huang, P.-F. Teng, C.-S. Lee, H.-L. Kwong, *Chem. Commun.* **2004**, 384–385.

133 C. Hamann, J.-M. Kern, J.-P. Sauvage, *Dalton Trans.* **2003**, 3770–3775.

134 M. D. Meijer, G. P. M. van Klink, G. van Koten, *Coord. Chem. Rev.* **2002**, *230*, 141–163.

135 J.-F. Nierengarten, *New J. Chem.* **2004**, *28*, 1177–1191.

136 D. Armspach, E. C. Constable, F. Diederich, C. E. Housecroft, J.-F. Nierengarten, *Chem. Commun.* **1996**, 2009–2010.

137 D. Armspach, E. C. Constable, F. Diederich, C. E. Housecroft, J.-F. Nierengarten, *Chem. Eur. J.* **1998**, *4*, 723–733.

138 U. S. Schubert, C. H. Weidl, A. Cattani, C. Eschbaumer, G. R. Newkome, E. He, E. Harth, K. Müllen, *Polym. Prepr.* **2000**, *41*, 229–230.

139 A. El-Ghayoury, A. P. H. J. Schenning, P. A. van Hal, C. H. Weidl, J. L. J. van Dongen, R. A. J. Janssen, U. S. Schubert, E. W. Meijer, *Thin Solid Films* **2002**, *403–404*, 97–101.

140 J. F. Nierengarten, J. F. Eckert, J. F. Nicoud, L. Ouali, V. Krasnikov, G. Hadziioannou, *Chem. Commun.* **1999**, 617–618.

141 C. Du, Y. Li, S. Wang, Z. Shi, S. Xiao, D. Zhu, *Synth. Met.* **2001**, *124*, 287–289.

142 S. Qu, C. Du, Y. Song, Y. Wang, Y. Gao, S. Liu, Y. Li, D. Zhu, *Chem. Phys. Lett.* **2002**, *356*, 403–408.

143 S. Qu, Y. Song, C. Du, Y. Wang, Y. Gao, S. Liu, Y. Li, D. Zhu, *Opt. Commun.* **2001**, *196*, 317–323.

144 R. Dagani, *Chem. Eng. News* **1999**, *77*, 54–59.

145 A. Khatyr, R. Ziessel, *Synthesis* **2001**, 1665–1670.

146 K. T. Potts, D. Konwar, *J. Org. Chem.* **1991**, *56*, 4815–4816.

147 S. Weidner, Z. Pikramenou, *Chem. Commun.* **1998**, 1473–1474.

148 J. M. Haider, M. Chavarot, S. Weidner, I. Sadler, R. M. Williams, L. De Cola, Z. Pikramenou, *Inorg. Chem.* **2001**, *40*, 3912–3921.

149 J. M. Haider, R. M. Williams, L. De Cola, Z. Pikramenou, *Angew. Chem. Int. Ed.* **2003**, *42*, 1830–1833.

150 U. Siemling, U. Vorfeld, B. Neumann, H.-G. Stammler, M. Fontani, P. Zanello, *J. Org. Met. Chem.* **2001**, 733–737.

151 J.-P. Collin, J.-O. Dalbavie, V. Heitz, J.-P. Sauvage, L. Flamigni, N. Armaroli, V. Balzani, F. Barigelletti, I. Montanari, *Bull. Soc. Chim. Fr.* **1996**, *133*, 749–754.

152 L. Flamigni, F. Barigelletti, N. Armaroli, B. Ventura, J.-P. Collin, J.-P. Sauvage, J. A. G. Williams, *Inorg. Chem.* **1999**, *38*, 661–667.

153 L. Flamigni, F. Barigelletti, N. Armaroli, J.-P. Collin, J.-P. Sauvage, J. A. G. Williams, *Chem. Eur. J.* **1998**, *4*, 1744–1754.

154 L. Flamigni, F. Barigelletti, N. Armaroli, J.-P. Collin, I. M. Dixon, J.-P. Sauvage, J. A. G. Williams, *Coord. Chem. Rev.* **1999**, 671–682.

155 H. T. Uyeda, Y. Zhao, K. Wostyn, I. Asselberghs, K. Clays, A. Persoons, M. J. Therien, *J. Am. Chem. Soc.* **2002**, *124*, 13806–13813.

156 L. Flamigni, I. M. Dixon, J.-P. Collin, J.-P. Sauvage, *Chem. Commun.* **2000**, 2479–2480.

157 E. Baranoff, J. P. Collin, L. Flamigni, J. P. Sauvage, *Chem. Soc. Rev.* **2004**, *33*, 147–155.

158 A. C. Benniston, G. M. Chapman, A. Harriman, M. Mehrabi, *J. Phys. Chem. A* **2004**, *108*, 9026–9036.

159 S. Schmatloch, C. H. Weidl, I. van Baal, J. Pahnke, U. S. Schubert, *Polym. Prepr.* **2002**, *43*, 684–685.

160 H. Hofmeier, A. El-Ghayoury, A. P. H. J. Schenning, U. S. Schubert, *Tetrahedron* **2004**, *60*, 6121–6128.

161 S. Ott, M. Kritikos, B. Åkermark, L. Sun, *Angew. Chem. Int. Ed.* **2003**, *42*, 3285–3288.

162 S. Ott, M. Borgstroem, M. Kritikos, R. Lomoth, J. Bergquist, B. Åkermark, L. Hammarstroem, L. Sun, *Inorg. Chem.* **2004**, *43*, 4683–4692.

163 K. M. Stewart, L. W. McLaughlin, *Chem. Commun.* **2003**, 2934–2935.

5

New Functional Polymers Incorporating Terpyridine Metal Complexes

5.1
Introduction

In today's polymer and materials research, a melding of tailor-made macro-molecules and traditional polymer chemistry has become an important feature of supramolecular systems. Thus, using a relatively large "classical" polymer content in the resulting supramolecular material, it has been possible to produce, on a large scale, materials combining the new and interesting features and structures of supramolecular species with the desired polymeric properties. Because of their polymeric nature, these materials can be, for example, spin-coated to form thin films or used in bulk, thus allowing technical processing via traditional polymer engineering methodologies. By altering the proportion of polymer, the macroscopic properties of the material can be dramatically changed, e.g., from rigid to elastic, from hydrophobic to hydrophilic, or from liquid to solid.

In addition to the well-known *H*-bonding systems [1], metal-ligand coordination is another important concept in supramolecular polymer chemistry. Chelate complexes of polypyridines, especially bi- and terpyridines of transition metals, have gained special attention recently because of their outstanding optical properties (see also Chapter 3).

Scientists started in the early 1970s to insert bipyridine units into large polyethylene glycolic macromolecules, ("crown ethers"), polymers, thin films, and membranes. The chemistry of 2,2':6',2"-terpyridines is – relatively speaking – much younger than that of their 2,2'-bipyridine counterparts. This also holds true for terpyridine-containing polymers, of which the first example appeared only recently [2]. This delayed inclusion is probably linked to the synthesis of the key functionalized terpyridines, which generally requires comparatively more steps than that of the 2,2'-bipyridine counterparts. Polymers containing bipyridine and phenanthroline moieties will not be considered here, but overviews on macromolecules bearing either phenanthrolines [3] or bipyridines [2, 4] are available. Other reviews concerning metal-containing polymers are also available [3, 5–7].

In the first part of this chapter, macromolecules bearing terpyridine units in the side chain will be discussed. These materials were the first terpyridine-

Modern Terpyridine Chemistry. U. S. Schubert, H. Hofmeier, G. R. Newkome
Copyright © 2006 WILEY-VCH Verlag GmbH & Co. KGaA, Weinheim
ISBN: 3-527-31475-X

containing polymers that were reported and were synthesized in a facile fashion by polymerizing monomers that were functionalized with appropriate terpyridine moieties. For a recent overview of dendronized polymers, Frauenrath [8] has addressed the different basic strategies. In the second part, polymers with terpyridine units in the main chain are described. By complexing *mono-* or *bis-* functionalized prepolymers, various architectures (e.g., extended linear or block copolymers) could be obtained [9]. Regarding biopolymer-terpyridine complexes, the focus to date is predominantly on macrobiomolecules including terpyridine complexes, which were covalently attached to enzymes, peptides or DNA/RNA. The supramolecular chemistry of simple terpyridine complexes with biosystems, such as DNA intercalators or ribozyme mimetics, has been considered in detail elsewhere [10, 11].

5.2
Polymers with Terpyridine Units in the Side Chain

To the best of our knowledge, Potts and Usifer were the first, in 1988, to incorporate the terpyridinyl unit into a polymer backbone [12] by (a) the free-radical homo-polymerization of either 4- or 4′-vinylterpyridine or (b) the copolymerized 4′-vinyl-terpyridine with styrene (Scheme 5.1). The resultant polymers were white powders with molecular masses of up to 60 000 g mol^{-1} (GPC with a polystyrene standard) and polydispersity indices in a range of 2–44. The addition of metal ions to these polymers generated an insoluble metallopolymer complex. The uncomplexed polymers could be recovered by using hot, concentrated hydrochloric acid. Although attempts to homopolymerize the corresponding vinylterpyridine metal complexes failed, styrene copolymers were readily formed from both [Co(4′-vinyl-tpy)(tpy)(PF$_6$)$_2$] and [Ru(4′-vinyl-tpy)$_2$(PF$_6$)$_2$] complexes [13].

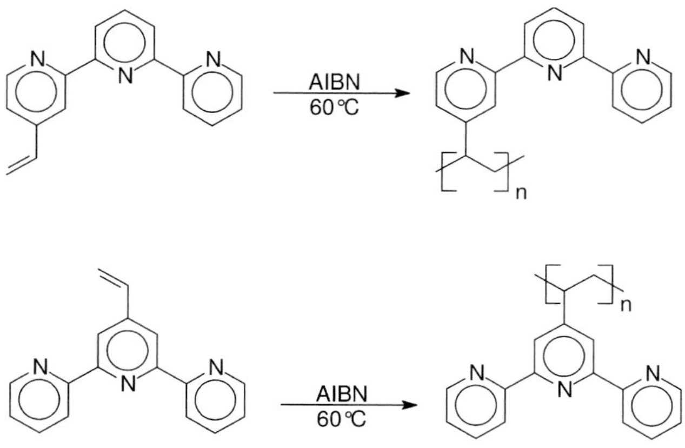

Scheme 5.1 The first reported terpyridine-containing polymers prepared by free-radical polymerization of vinylterpyridines [12].

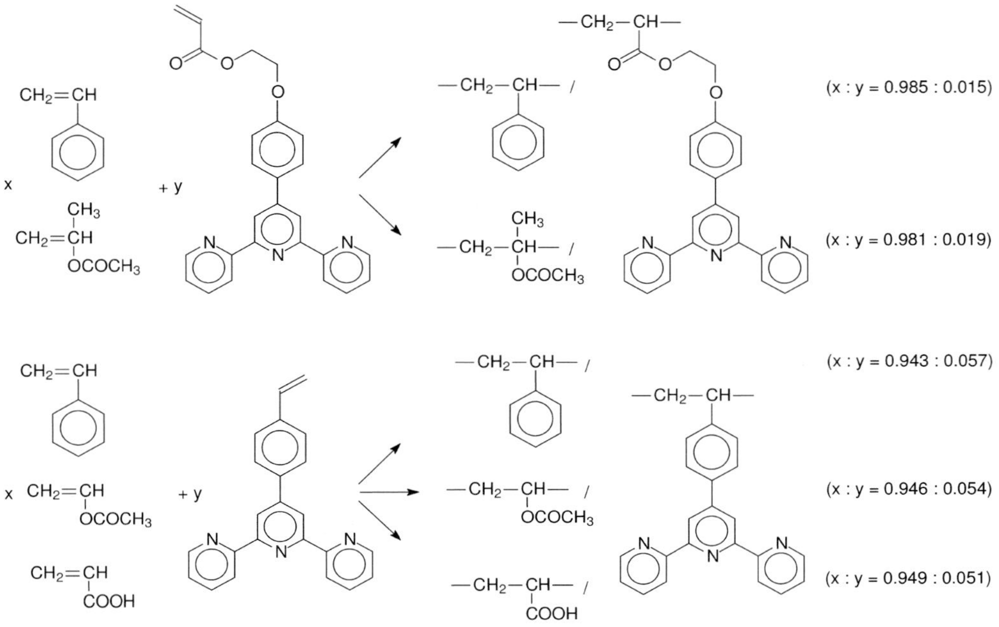

Scheme 5.2 Copolymers with pendant terpyridine units in the side chain [14, 15].

Later in 1990 and 1992, Hanabusa et al. published [14, 15] an extended series based on either 4'-[4-(2-acryloyloxyethoxy)phenyl]terpyridine or 4'-(4-styryl)-terpyridine as monomers for homo- and copolymerization (Scheme 5.2). In the former, the copolymerization was conducted using styrene or methyl methacrylate, producing an average molecular weight (GPC) of 6400 g mol^{-1} for the styrene copolymer and 15 000 g mol^{-1} for the copolymer with methyl methacrylate. The latter paper described the homopolymers of 4'-(4-styryl)terpyridine, and copolymers with styrene, vinyl acetate, and acrylic acid. While these homopolymers were insoluble, all three copolymers were, however, soluble, because of their low terpyridine content (5%). In all cases, the corresponding metal complexes were formed with CoCl$_2$, FeCl$_2$, NiCl$_2$, and CuCl$_2$.

Recently, Tew et al. prepared a random copolymer using a methyl methacrylate that was functionalized with terpyridine, as the co-monomer (Figure 5.1, left) [16]. Upon addition of Cu(II) nitrate, an increase in viscosity was observed, which did not, however, occur in the case of the related homopolymer. Even at low concentrations of 4 mg mL^{-1}, a significant rise in viscosity was observed, which, when compared to other non-covalent systems such as *H*-bonding, had not been observed [17].

A very similar approach to such copolymers was recently published by Schubert and Hofmeier in which the only difference was the spacer length between the 4'-position of the terpyridine and the polymer backbone [19]. Detailed viscosity studies

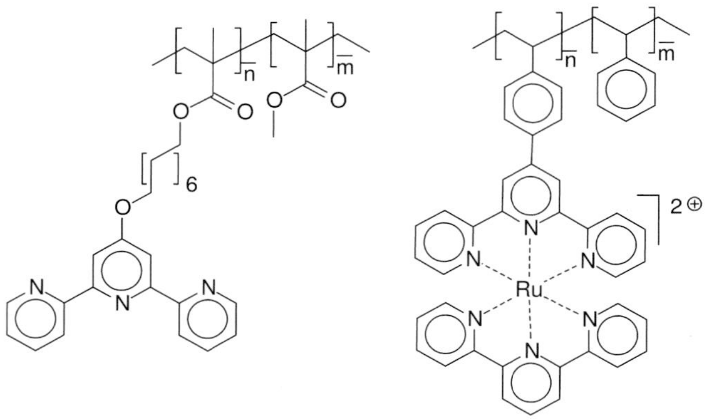

Figure 5.1 Left: Tew's methacrylate copolymer for metal complexation [16]. Right: Heller's Ru-coordinated polystyrene [18].

through stepwise addition of Fe(II) or Zn(II) ions were performed showing that the viscosity dependence was based on terpyridine content as well as concentration, thus providing evidence for the formation of large aggregates of different sizes (Figure 5.2) [20].

Moreover, the type of metal ions can also play an important role; for example, the addition of Zn(II) ions led to $[Zn(R\text{-}tpy)_2]^{2+}$ cross-linked metallopolymers possessing a lower viscosity. After an overtitration, the relative viscosity eventually dropped because these Zn(II) complexes are reversible and can also form *mono*-complexes, $[Zn(R\text{-}tpy)(X)_2]$, in the presence of an excess of Zn(II) ions. At even higher Zn(II) ion concentration, gel formation was observed due to the generation of an extended network [21]. The terpyridine copolymer was then used to create a more complex metallopolymer. With the traditional Ru(III)/Ru(II) method, graft copolymers were synthesized in which additional $[Ru(4'\text{-}PEG\text{-}tpy)(X)_3]$ possessing an end-functionalized poly(ethylene glycol) chain could be attached to the terpyridines located on the polymer backbone via $[Ru(tpy)_2]^{2+}$ connectivity. Additionally, an $[Ru(4'\text{-}polylactide\text{-}tpy)(X)_3]$ and a non-polymeric $[Ru(4'\text{-}R\text{-}tpy)(X)_3]$ were used as grafting ligand complexes (Scheme 5.3). When comparing the thermal properties of the grafted copolymers with the starting copolymer, differential scanning calorimetry (DSC) measurements showed significant differences, especially in the glass transition temperatures, thus demonstrating the influence of the grafting process.

A different strategy toward side-chain, terpyridine-functionalized polymers utilized commercial polymers; for example, Schubert et al. modified polyvinyl-chloride (PVC) by an initial treatment with (2-mercaptophenyl)methanol to introduce hydroxy groups, and then reaction with a terpyridine bearing an isocyanate group (Scheme 5.4). The resulting terpyridine-modified polymer was investigated regarding grafting with an *oligo*(ethylene glycol) and cross-linking through addition of Fe(II) ions [22].

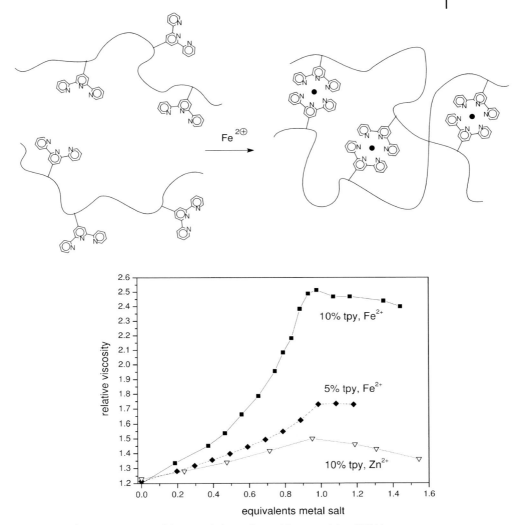

Figure 5.2 Schematic overview of the cross-linking of terpyridine-containing PMMA and viscosity plots of the stepwise addition of metal salts to the copolymer [20].

Recently, an example of a *divergent* approach to polymerizing $[M(tpy)_2]^{2+}$ complexes, which are functionalized on one side with a polymerizable group, was reported (Figure 5.1, right) [18]. Hetero-$[Ru(4'-CH_2=CH-tpy)(4'-HOCH_2-tpy)]^{2+}$ complexes were copolymerized with styrene using simple free-radical polymerization; the product precipitated as an orange solid, which had an M_n (determined by GPC) of 5170 with a polydispersity index (PDI) of 1.62. The fact that no dissociation was observed when it was passed through the GPC column also demonstrated the stability of such hetero-$[Ru(R-tpy)(R'-tpy)]^{2+}$ complexes. Through

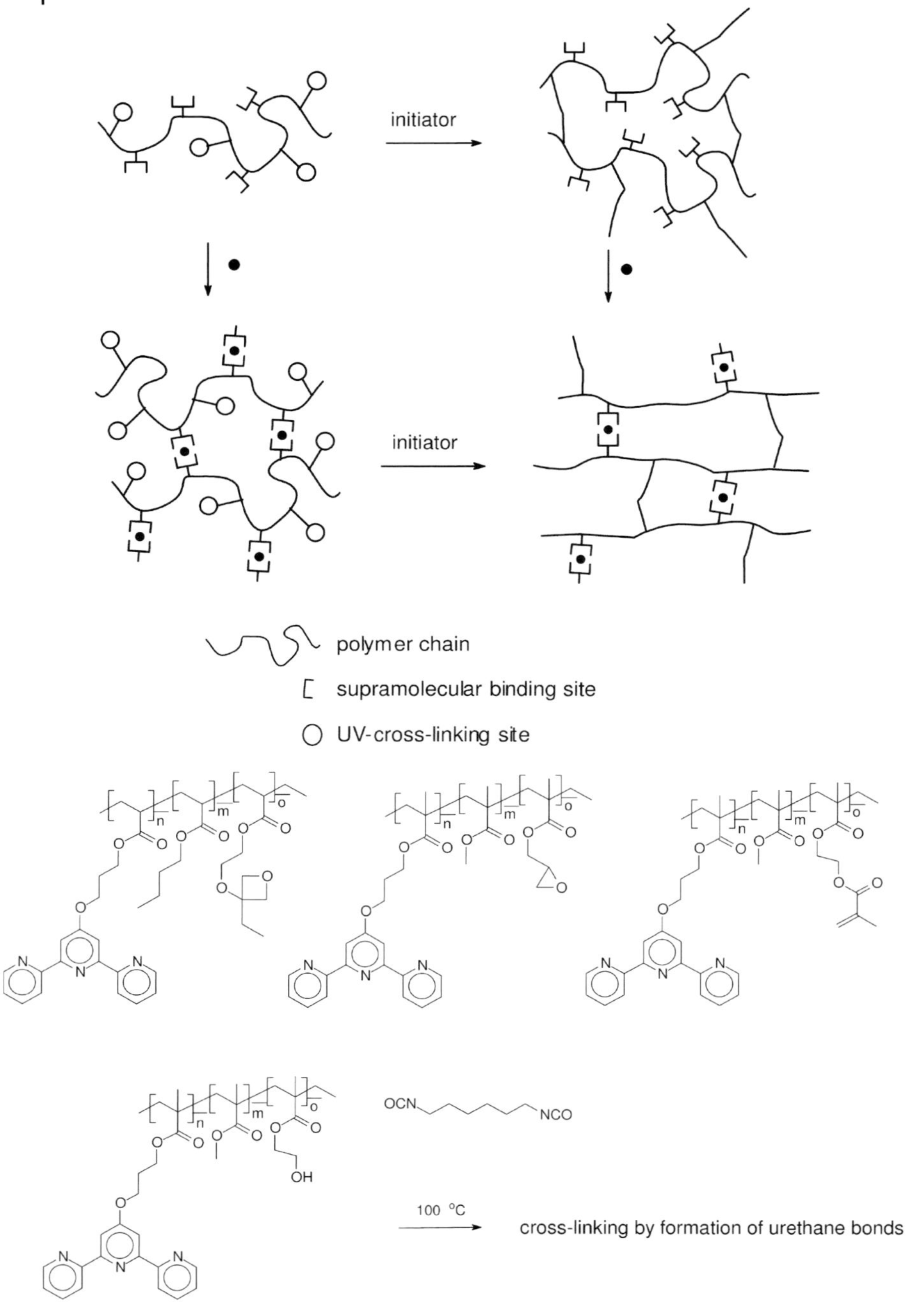

Scheme 5.5 Top: overview of the two-step covalent and supramolecular cross-linking; bottom: synthesis of the terpolymer.
(Reprinted with permission from [24], © 2003 American Chemical Society).

An attempt has been made to combine and individually address $[M(tpy)(tpy')]^{2+}$ complexation and the UV-mediated polymerization of oxetanes [24]. Therefore, an acrylate terpolymer was designed bearing a terpyridine as the metal-complexing unit and an oxetane as a covalent cross-linking unit (Scheme 5.5). The terpolymer obtained through radical copolymerization had an M_n of 7400 g mol^{-1} and an average terpyridine content of 2.8 per polymer chain. It was shown that the addition of FeCl$_2$ led to the typical MLCT absorption (purple coloring) of the $[Fe(tpy)(tpy')]^{2+}$ complex. Upon subsequent reaction with AlCl$_3$, ring-opening and covalent cross-linking of the oxetane groups occurred. The order of the cross-linking could also be reversed, in which first a rubber-like material was obtained from the covalent cross-linking by treatment with AlCl$_3$. Subsequent treatment of the cross-linked polymer with a methanolic solution of FeCl$_2$ immediately resulted in a purple coloration, indicating that additional coordinative cross-linking had occurred. This demonstrated the feasibility of such $[M(tpy)(tpy')]^{2+}$ complexation as part of a multi-step, cross-linking procedure, which might be of interest, for example, for smart coatings with self-healing properties. Subsequently, this approach was extended to terpolymers containing epoxide moieties [21]. In addition to analogous oxetane studies, the cross-linking density through gel swelling studies was investigated. The material, when subjected to Fe(II) ions as well as AlCl$_3$, revealed a higher degree-of-cross-linking than that in cases where only either covalent or supra-molecular cross-links were present, which is consistent with both types of cross-linking. However, the terpolymers containing epoxide or oxetane groups showed minimal efficiency toward UV-initiated curing. This is probably caused by an acid-base reaction of the photoinitiator with a terpyridine moiety. Therefore, terpolymers were designed possessing either (a) (meth)acrylates for free-radical, UV-curing or (b) hydroxy groups that could be thermally cured with isocyanates [25].

In 1998, Kimura, Hanabusa, et al. introduced a fluorescing poly(*p*-phenylene-vinylene) with pendant terpyridinyl groups, as a chemosensor for metal ions [26]. As shown in Scheme 5.6, polymerization was conducted by a Wittig-type reaction of 2,5-*bis*(hexyloxy)benzene-1,4-dialdehyde and the terpyridinyl-phosphonium salt, yielding a polymer with $\bar{M}_w = 4000$, as determined by GPC (polystyrene standard). The fluorescence of the polymer was investigated as a function of various metal ions; at 524 nm, the fluorescence was completely quenched by Fe(II), Fe(III),

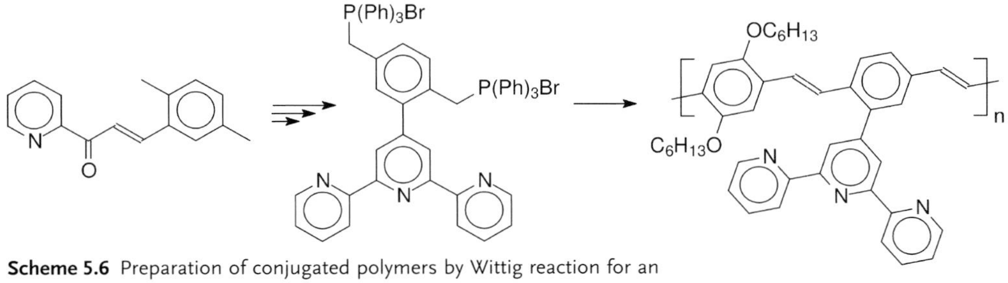

Scheme 5.6 Preparation of conjugated polymers by Wittig reaction for an application as fluorescing chemosensors [26].

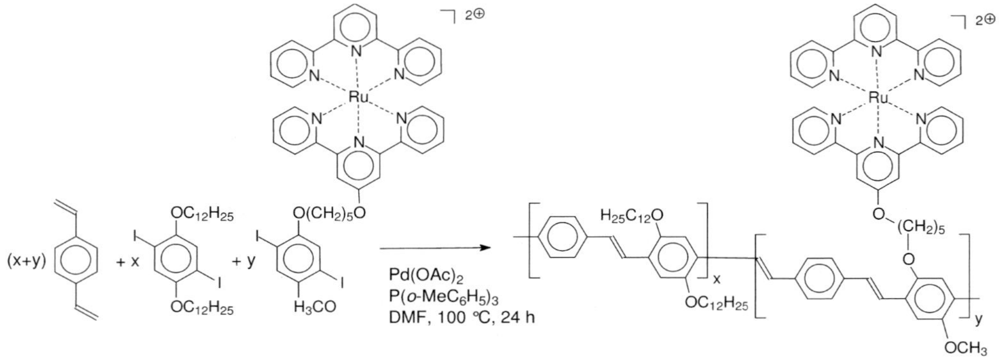

Scheme 5.7 Light-emitting polymers with a pendant [Ru(4′-O-tpy)(tpy)]$^{2+}$ complex [27].

Ni(II), Cu(II), Cr(II), Mn(II), and Co(II). A blue shift in the emission spectrum was shown for Pd(II), Sn(II), Al(III), and Ru(II).

A poly(*p*-phenylenevinylene) (L) with pendant [Ru(4′-RO-tpy)(tpy)]$^{2+}$ and [bipyridin-4-yl]-Ru(II) complexes has recently been synthesized by Chan et al. [27]. In this synthesis, 1,4-divinylbenzene, 1,4-di(dodecanoxy)-2,5-diiodobenzene, and an [Ru(4′-I-tpy)(tpy)]$^{2+}$ complex were copolymerized in various ratios via a Heck coupling procedure (Scheme 5.7). These polymers exhibited photo- and electroluminescence at room temperature. Photocurrent measurements were conducted on samples that were spin-coated onto an ITO surface at 490 nm, demonstrating photoconductivities in the order of magnitude of $10^{-12}\ \Omega^{-1}\ cm^{-1}$.

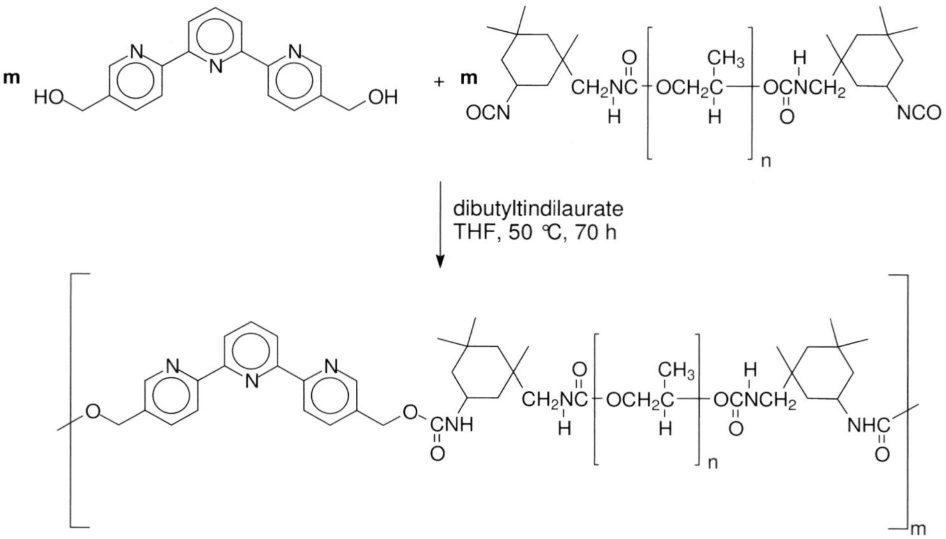

Scheme 5.8 Copolymerization of *bis*(hydroxymethyl)terpyridine with a *bis*-isocyanate-functionalized prepolymer.

LEDs produced from these polymers exhibited a turn-on voltage of 5 V with rectification greater than 10^3 at 15 V and a maximum luminance of 360 cd/m^2.

In a different approach, a 5,5″-*bis*(hydroxymethyl)terpyridine was reacted with a commercially available *bis*isocyanate-functionalized prepolymer, resulting in an AB multiblock copolymer (Scheme 5.8) [28]. Subsequent treatment of a solution of the block copolymer with Co(OAc)$_2$ in chloroform immediately revealed a red-brown coloration. In contrast to the viscous free polymer, the complexed metallo-product had rubber-like properties.

5.2.1
Polymers with Terpyridine Units in the Polymer Backbone

The combined properties of conventional polymers with those of [M(tpy)(tpy′)]$^{2+}$ complexes have become of increasing interest over the last few years. There are mainly two different chemical approaches to introducing terpyridines and their complexes into the backbone polymeric systems: (a) by functionalizing properly modified polymers with terpyridine ligands or (b) by using a functionalized terpyridine as an initiator (the *convergent* approach starting from uncomplexed terpyridine). These main approaches also apply to corresponding *bis*terpyridine metal complexes (*divergent* approach, in which metallopolymers are formed starting from the complex). Having a functionalized polymer with free terpyridine ligands and subsequently forming different combinations of *bis*-complexes with different metals leads to a rich variety of possible new metallosuperstructures. For a detailed insight into these strategies, the reader should see recent overviews [2, 29, 30].

These polymeric terpyridine materials can be either *mono*functionalized or of a telechelic nature, possessing two or more terpyridines per chain. Having ter-pyridine units at both ends of each chain allows access to linearly extended chains containing metal "linkers". Scheme 5.9 shows the concept of such a polymer.

In 1995, Constable discussed his own work exploring the concept of using metallo-supramolecular principles to prepare oligomers and polymers with pre-coded properties by coordination to metal ions [31].

Although the successful synthesis of a coordination polymer has not yet been reported, the structures shown here give an impression of the true potential of this strategy with respect to the incorporation of distinctive physical properties. Some groups have investigated rigid rods during their research, addressing long-distance interaction phenomena, such as electron and energy transfer or magnetic

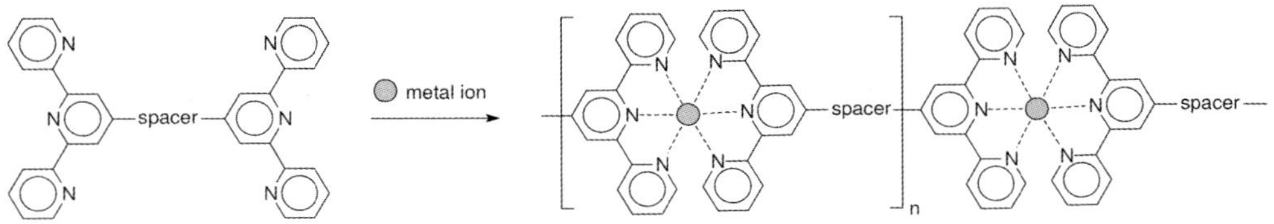

Scheme 5.9 Polymeric *bis*terpyridine-metal complex (charge and anions omitted).

coupling of transition metal ions (see "dyads" and "triads" in an earlier chapter) [32]. Although no polymers were formed, structures were obtained that could act as very interesting building blocks for non-covalent polymers because of their *bis*terpyridine functionalization.

5.2.2
Polymers from Rigid Organic Building Blocks

Möhwald, Kurth et al. used analytical ultracentrifugation for the characterization of a coordinating polymer built of 1,4-*bis*(terpyridin-4′-yl)benzene (Scheme 5.10), which was first introduced by Constable et al. [33], and treated it with Fe(II) ions [34]. Applying the Svedberg formula, they determined a minimum molecular mass of 14 900 g mol^{-1}, corresponding to 25 repeating units; however, this can only be considered a rough estimate because of the uncertainty in the determination of its partial specific volume $\bar{v}$ [35]. Interesting layer-by-layer self-assembly with this polymer was conducted by Kurth et al. The resulting alternating monolayers were investigated by X-ray reflectivity measurements, XPS, and surface plasmon resonance spectroscopy [36]. A thickness of dried polymer films of 18 ± 3 Å was determined, corresponding to a single layer (see also results utilizing isolated terpyridine metal complexes) [37–41].

In a recent paper, Kurth et al. produced anisotropic thin-film materials of this metallo-supramolecular polymer by electrostatic binding to the amphiphile

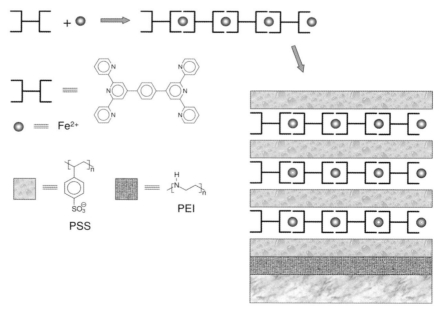

Scheme 5.10 Linear metallo-supramolecular coordination polyelectrolytes used for the formation of a superlattice by layer-by-layer self-assembly [36].

dihexadecyl phosphate [42]. The resulting polyelectrolyte-amphiphile complex formed a stable monolayer at the air-water interface that is readily transferred and oriented on solid supports by the Langmuir-Blodgett technique. The average thickness per layer was demonstrated to be 2.8 ± 0.2 nm. The presented strategy opens up a new route to materials with tailored structures and functions. Kurth et al. have demonstrated simple methods for the fabrication of devices with metallosupramolecular substrates [43–47]. Recently, they created a polyelectrolyte by the reaction of Co(II) with a *bis*terpyridine ligand incorporating a bridging metalloviologen unit; the fabrication of thin-multilayers via layer-by-layer self-assembly generated an electrochromic film [44].

The same ligand has been used by Mohler et al. for the controlled stepwise self-assembly of rigid rods [48]. Utilizing ruthenium chemistry, these authors built up a well-defined linear oligomer consisting of seven metal centers (Scheme 5.11). Electrospray mass spectrometry was applied to characterize a solution of the heptanuclear complex in MeCN.

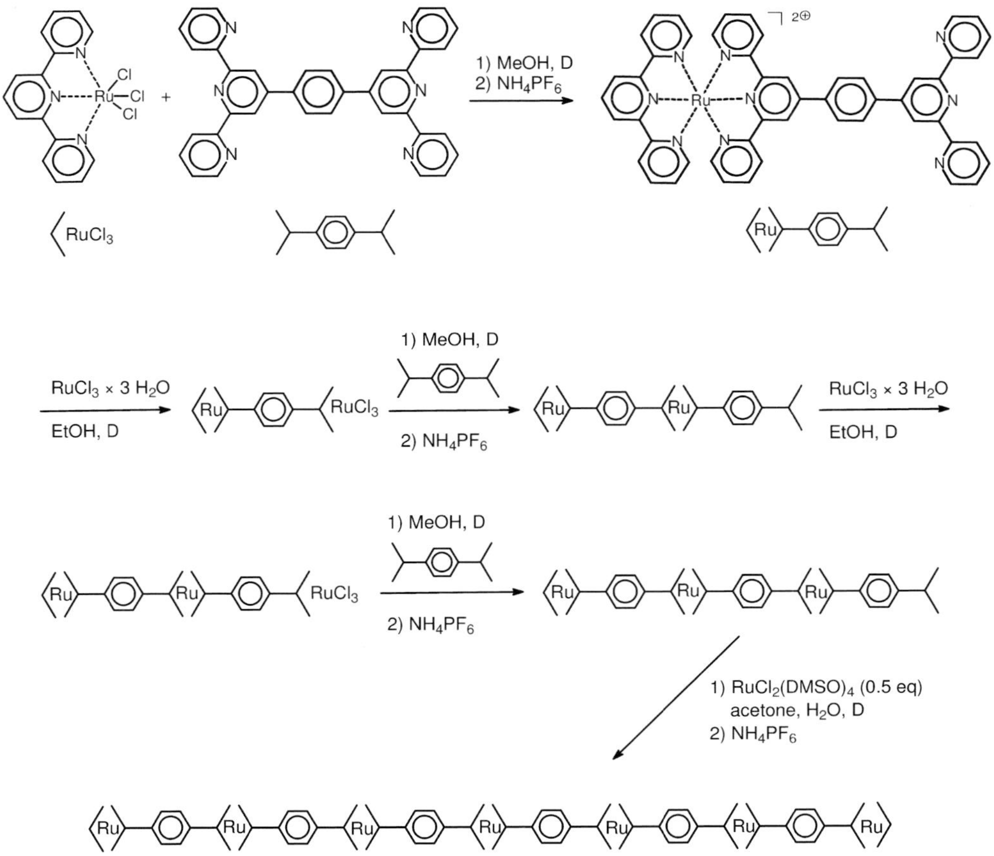

Scheme 5.11 Well-defined linear rods consisting of seven metal centers [48].

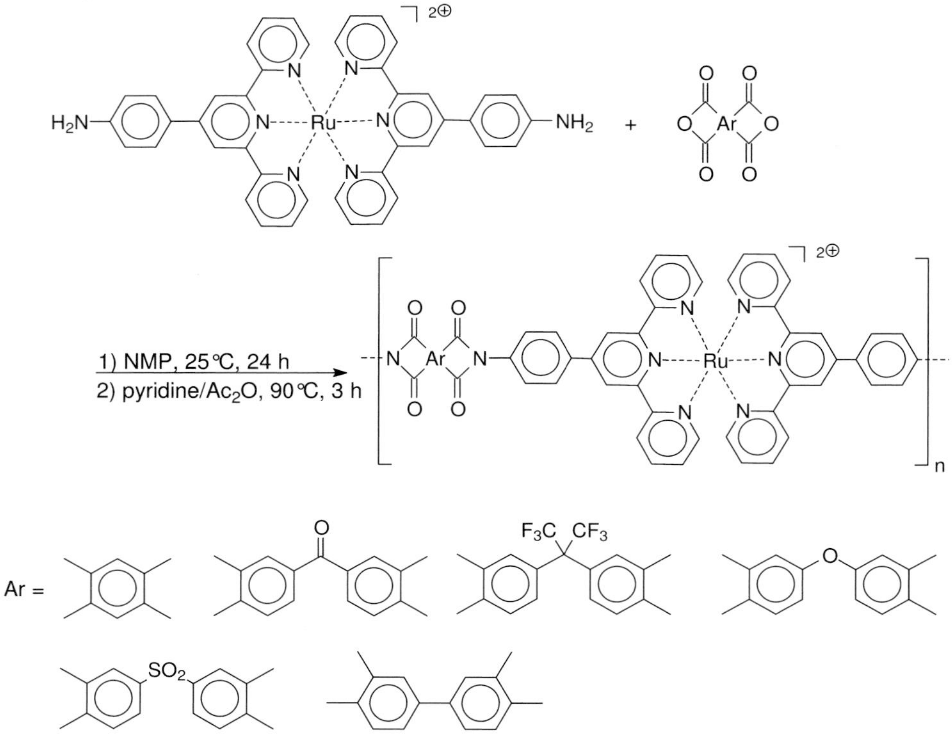

Scheme 5.12 Novel aromatic polyimides for application as LEDs [49–51].

A series of poly(p-phenylenevinylene) polymers and aromatic polyimides (Scheme 5.12) have been synthesized and characterized by Chan et al. [49–52]. To generate poly(p-phenylenevinylene)s, 1,4-divinylbenzene, 1,4-dibromo-2,5-di(hexyloxy)benzene and $[Ru(4'-I-C_6H_4-tpy)_2]^{2+}$ were coupled in different ratios by Heck coupling procedures, resulting in polymers with the terpyridine complex incorporated into the backbone [49]. By varying the monomer ratios, the macroscopic properties, e.g., solubility, of the polymers changed. The polyimides were copolymerized from an $[Ru(4'-H_2N-C_6H_4-tpy)_2(PF_6)_2]$ complex with different aromatic dianhydride monomers [50]. The electron and hole carrier mobilities of the polyimides were demonstrated to be within an order of magnitude of 10^{-4} cm^{2}V^{-1}s^{-1}. Single-layered light-emitting diodes were produced with an external quantum efficiency of 0.1% and a maximum luminescence of 120 cd m^{-2}. Interestingly, Chan et al. in both cases used this rather traditional approach with preformed complexes for polymerization only, not the coordinative polymerization of *bis*nucleating ligands by addition of metal ions.

Up to now, the most extensive study of terpyridine-based metal coordination polymers has been performed by Kelch and Rehahn in 1999 [53, 54]. To enhance solubility, the rigid ligand was functionalized with two n-hexyl groups (Scheme

5.13). From end-capping experiments combined with NMR investigations, the degree-of-polymerization was demonstrated to be at least 30, corresponding to a molecular weight of more than 36 000 g mol^{-1}. These high values obtained from NMR were further supported by viscosity measurements. The intrinsic viscosity [η] was determined to be ca. 300 mL g^{-1}, which is of the same order of magnitude as the intrinsic viscosities observed for poly(*p*-phenylene)s. A strong polyelectrolyte effect due to the charges in the backbone was also found for these polymers, but could, however, be suppressed by performing the viscosity measurements in solutions in which a screening salt was added.

In a more recent contribution, analogous rod-like [Ru(tpy)$_2$]$^{2+}$ polymers were prepared in which a heteroleptic complexation was used, where one of the ligands does not consist of a terpyridine, but rather a 2-phenylbipyridine. In addition to the 5 nitrogens, a carbanion completes the complexation at the Ru(II) center. From ^{1}H NMR end-group analysis, it was concluded that these Ru(II) complex polymer chains have a DP ≥ 20, which corresponds to the previous results (Scheme 5.14) [55].

The above approaches utilizing long side chains to yield soluble rod-like non-covalent polymers in solution have a notable drawback, i.e. they are derived from rather complicated building blocks as well as coupling procedures.

Meijer et al. reported the synthesis of a rigid [Fe(tpy)$_2$]$^{2+}$ polymer including oligo(phenylene vinylene) (OPV) units (Figure 5.4, top) [56]. Because of this rigidity, the formation of small macrocycles is unlikely, and the degree-of-polymerization was estimated to be DP ≈ 100 at the applied millimolar concentration derived from kinetic data obtained from a UV/Vis titration experiment. Such metallo-polymers containing Ru(II) as the metal center are of special photophysical interest because of their fluorescence (even at room temperature) provided by the conjugated system attached at the 4′-position.

Similar systems containing a chiral pinene moiety attached to the outer terpyridine rings were reported by Barron et al. (Figure 5.4, bottom) [57]. Extensive luminescence lifetime studies were conducted on these systems, demonstrating that the length of the π-electron delocalization "box" governs the emission energy at 77 K, where the luminescence lifetimes are not controlled by non-radiative decay, in contrast to the room-temperature experiments.

A further example using the different approach of first synthesizing a 4′-functio-nalized [Ru(tpy)$_2$]$^{2+}$ complex and subsequent polymerization of this complex to a ruthenium complex polymer was reported [58] by Constable et al., in which thiophene moieties, attached to the 4′-positions of the terpyridines in the Ru(II) complex, were electrochemically polymerized (Scheme 5.15). From semi-empirical calculations (ZINDO/S), it was concluded that, because of the build-up of a positive charge at the thienyl C5 position, that position is the favored one for coupling to give a linear, rod-like Ru(II)-complex polymer. The authors did not determine the degree-of-polymerization, but did, however, investigate the deep red film (100 nm thick) formed on the electrode surface using AFM and XPS (detection of ruthenium 3d$_{5/2}$ and S 2p$_{3/2}$). Furthermore, the red shift of the MLCT-band from 498 to 521 nm matches the extended conjugation arising from the bithienyl spacer.

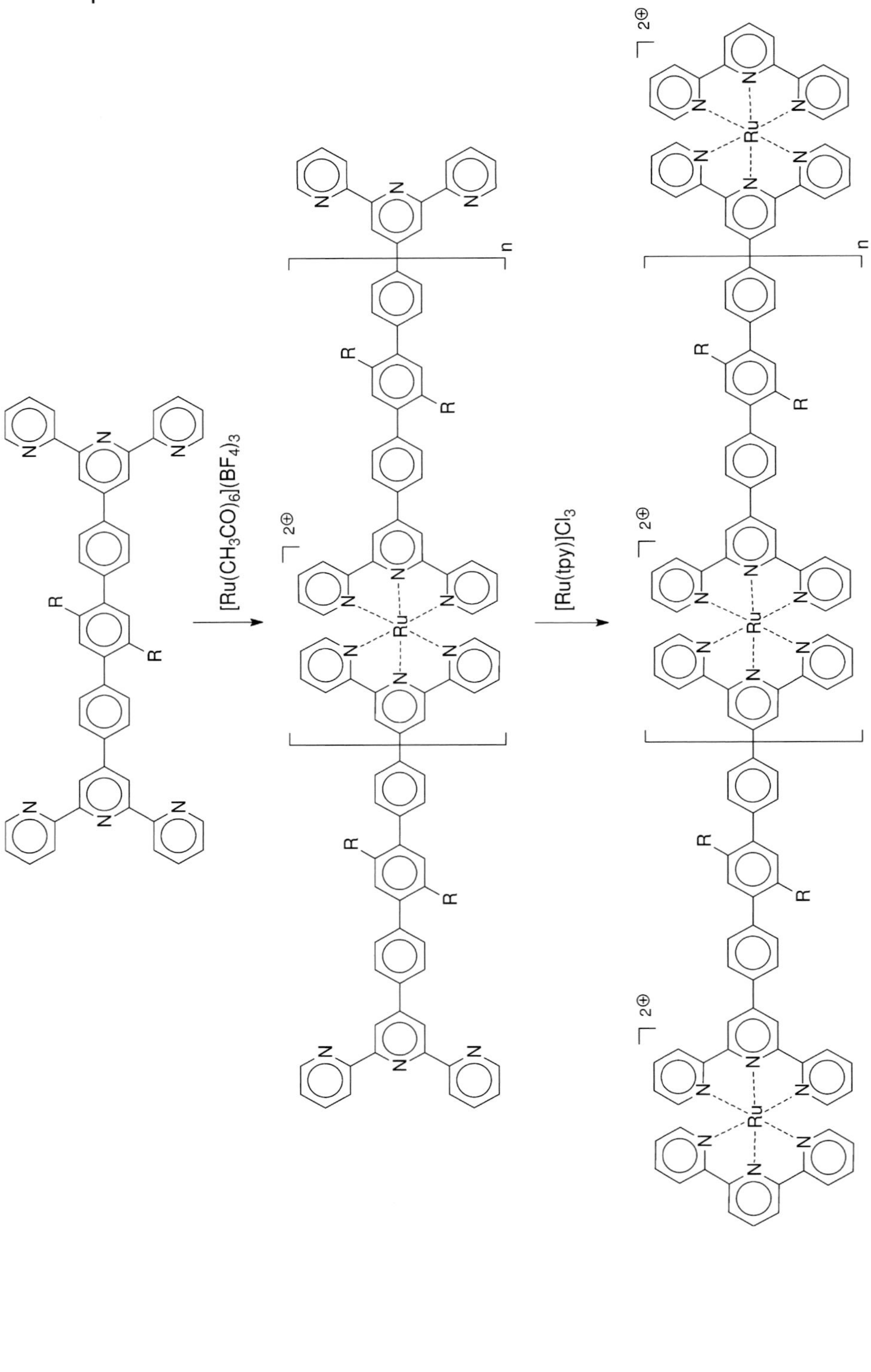

Scheme 5.13 Rigid rod-like polymer based on [Ru(tpy)$_2$]$^{2+}$ complexes [53, 54].

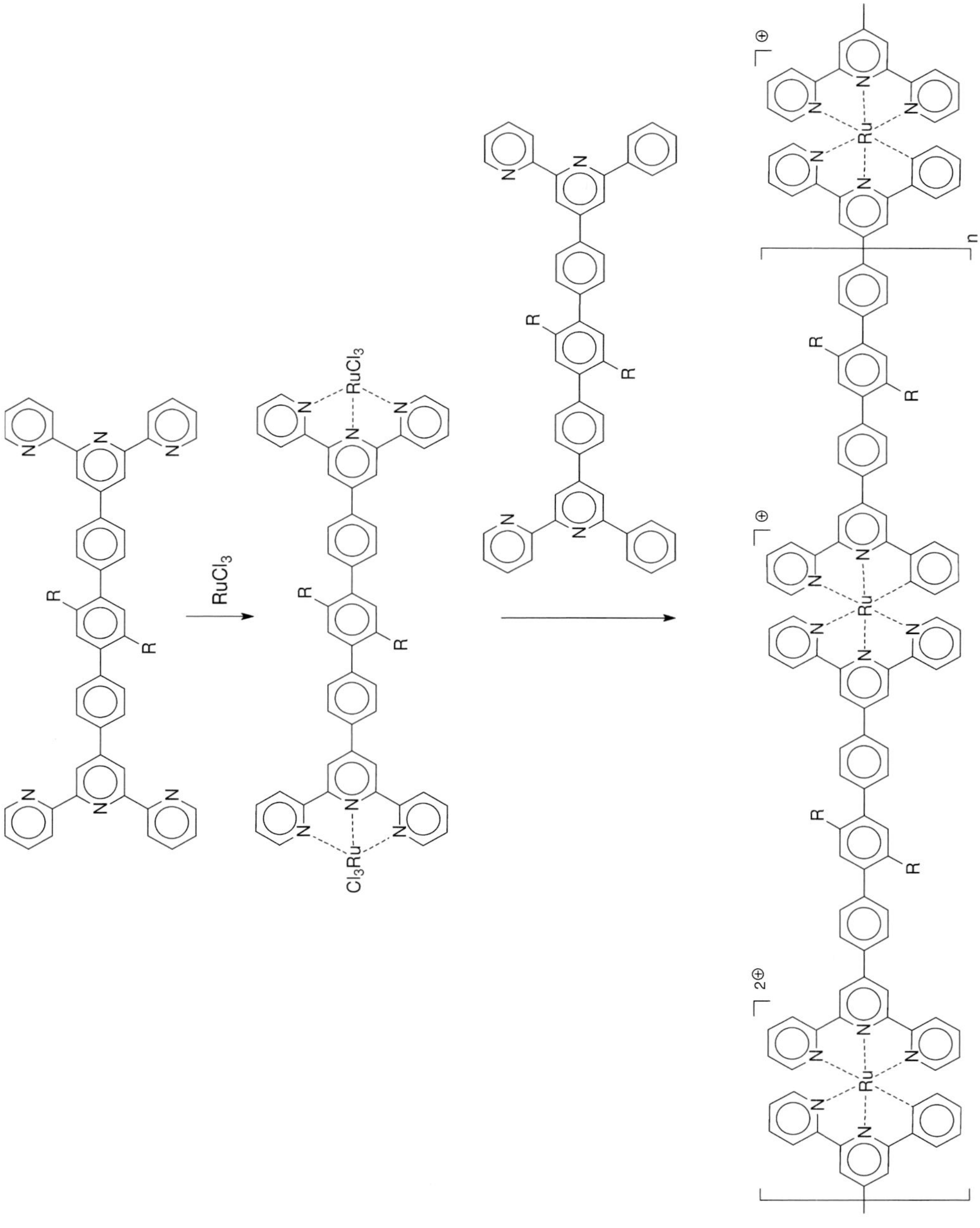

Scheme 5.14 Rehahn's synthetic route selected for the preparation of a [Ru(tpy)₂]²⁺ coordination polymer [55].

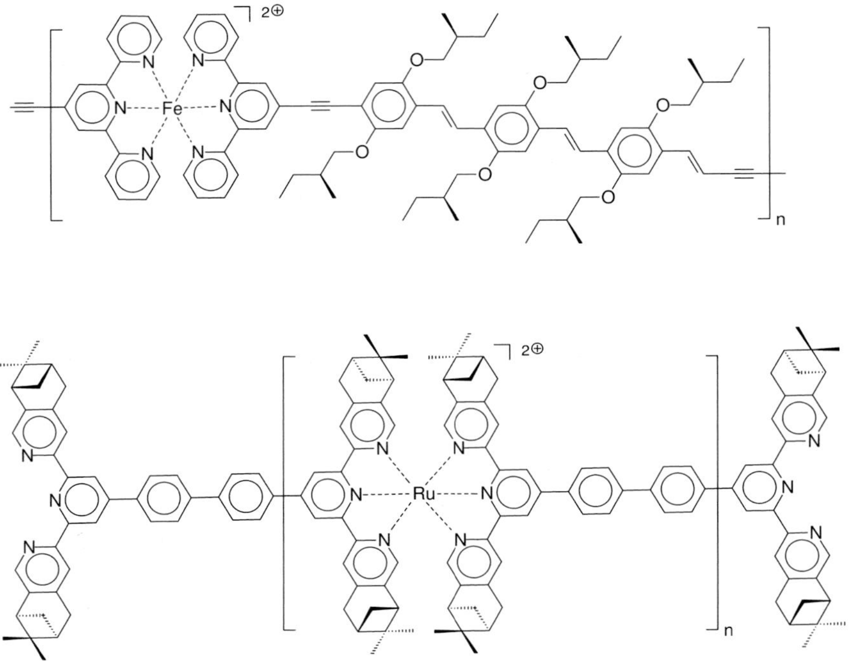

Figure 5.4 Top: combination of an $[Fe(tpy)_2]^{2+}$ complex and an oligo(phenylenevinylene) unit [56]. Bottom: chiral $[Ru(tpy)_2]^{2+}$ complex polymer [57].

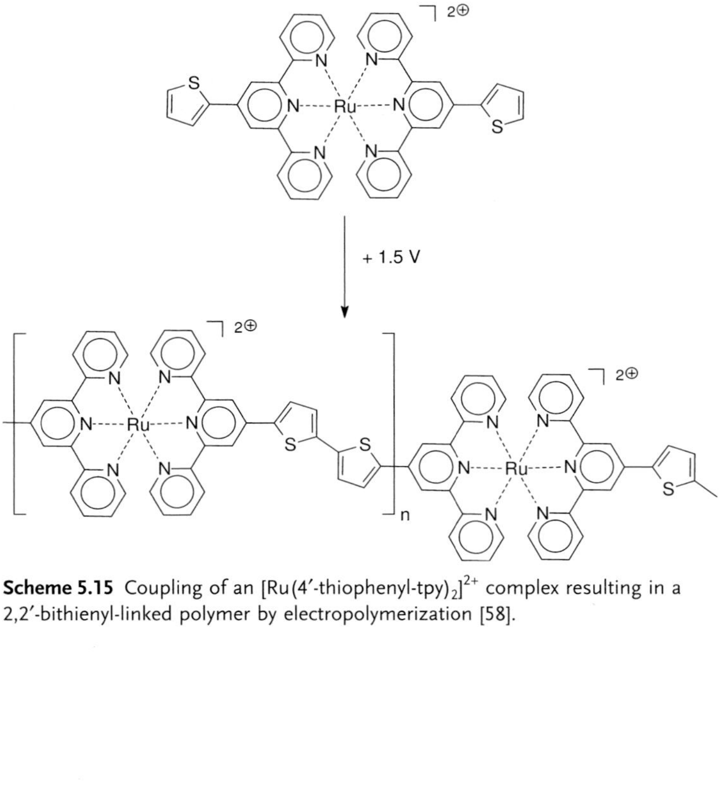

Scheme 5.15 Coupling of an $[Ru(4'-thiophenyl-tpy)_2]^{2+}$ complex resulting in a 2,2'-bithienyl-linked polymer by electropolymerization [58].

5.2.3
Polymers from Flexible Organic Building Blocks

In the previous section, supramolecular polymers from rigid precursors were described and shown to possess rod-like structures. Since most polymers used for industrial applications are flexible macromolecules, flexible linkers have been introduced into the precursors for supramolecular polymers.

A series of *bis*-terpyridine-functionalized ligands as well as the corresponding polymeric species have been reported by Colbran et. al. [59, 60]. Starting with 4'-(4-aminophenyl)terpyridine, polymers were prepared by two different synthetic routes: (a) coupling of two ligands with difunctional organic reagents, e.g., with pyromellitic anhydride, terephthaloyl chloride, or adipoyl chloride, and then treating the new binucleating ligands (Figure 5.5) with metal ions, or (b) pre-forming monomeric complexes of the 4'-(4-aminophenyl)terpyridine followed by treatment with these same difunctional organic reagents. In approach (b), the polymerization was quenched and a molecular weight of 18 kDa, corresponding to a degree-of-polymerization of 17, was calculated based on the number of end groups. Similar results were obtained for the polymers resulting from approach (a).

Recently, there have been numerous different approaches in this direction, mainly focusing on Fe(II) and Ru(II), as the "metal glue" for coordination polymerization. *Bis*terpyridine complexes of these two metals possess very high

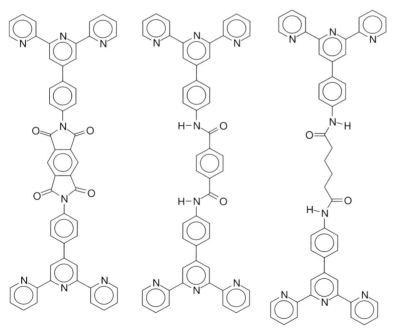

Figure 5.5 Organic ligands suitable for transition metal ion-based non-covalent polymerization [59, 60].

stability constants, and only in the case of Fe(II) is reversibility known to occur upon heating [61]. Recently, Kimura et al. [62] reported a chiral example obtained from twisted enantiomeric bridging ligands (Figure 5.6). The enantiomeric ligands, consisting of a *bis*-terpyridine-functionalized chiral binaphthyl spacer, are separately complexed with one equivalent of FeCl$_2$ in order to yield the enantiomeric coordination polymers. Upon addition of up to one equivalent of FeCl$_2$, a nearly linear increase in UV/Vis absorption of the Metal-to-Ligand Charge-Transfer (MLCT) band could be obtained. Moreover, circular dichroism (CD) spectroscopy showed an increase in the MLCT band as well, suggesting a chiral induction to the optically inactive [Fe(tpy)$_2$]$^{2+}$ moiety and therefore superstructural chirality; however, characterization of these novel polymeric or oligomeric species is, needless to say, challenging, in that data concerning polydispersity index or ring-chain equilibrium are difficult to obtain.

Abruña et al. reported the synthesis of bridging *bis*-terpyridine ligands which possess inherent chirality [63]. Complexation with Fe(II) salts was studied in solution with UV/Vis and CD spectroscopy in order to confirm the construction of chiral metallo-assemblies. From the UV/Vis titration with Fe(BF$_4$)$_2$, it was shown that the isolated MLCT band increased in intensity up to an equivalent of 1 : 1, but remained unchanged even after over-titration (Figure 5.7). This proves, as reported by Kimura et al., that formation of saturated octahedral *bis*-complexes can be confirmed by CD spectrometry with additional evidence for the formation of an optically active compound upon complexation with Fe(II). ESI-MS measurements showed that the polymeric assemblies undergo severe fragmentation under these conditions; however, large fragments with up to 9 ligand-Fe(II) repeating units were still observed.

Constable et al. recently reported the separation by column chromatography of different fractions from their polymerization attempts using a *bis*-terpyridine-tri(ethylene glycol) and FeCl$_2$. There is evidence, according to electrospray mass spectrometry and ^{1}H NMR experiments, that the two main products formed are 3+3 and 4+4 macrocycles (see Section 4.4 "Cycles" in previous chapter) [64]. Only a smaller fraction, which could not be eluted from the silica column, is believed to be composed of large macrocycles or linear polymers. Similar observations were made by Schubert et al. using hexane instead of tri(ethylene glycol) as the spacer. In addition, it was shown in this latter case that it is possible to influence macrocycle formation by applying heat to the mixture after performing the metallo-polymerization at room temperature [65]. According to ^{1}H NMR and MALDI-TOF-MS results, numerous different components were observed. Analogous results were obtained for a *bis*-terpyridine containing a hexadecyl spacer, where species possessing up to 9 monomers could be detected by MALDI-TOF-MS. This telechelic component showed good solubility, thus permitting a viscosimetric titration with Fe(II) ions; a significant rise in the solution viscosity indicated the formation of extended polymers [66]. It is proposed that entropy-driven formation of smaller cycles from large polymeric species occurs on exchange after heating, which is comparable to Constable's results. (Generally, exchange upon heating is known to occur for [Fe(tpy)$_2$]$^{2+}$ systems [61].) For another similar metallopolymer

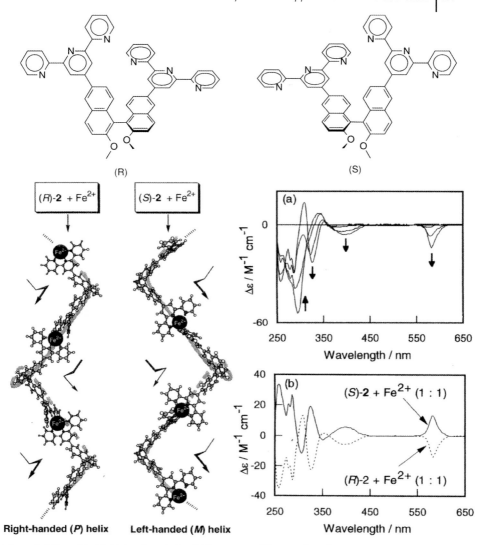

Figure 5.6 Top: chiral bridging ligands in which two terpyridine exo-ligands are linked at 6- and 6′-positions by a chiral binaphthyl spacer. Bottom left: schematic representation of the stereospecific assemblies of Fe(II) and (R)- and (S)-chiral bridging ligands. Bottom right: (a) effect of Fe(II) concentration on the CD spectra of the (R) enantiomer (0.1 mM) in CHCl$_3$-MeOH at 20 °C: [Fe(II)] = 0, 0.02, 0.06, 0.1 mM. Arrows indicate the spectral change. (b) CD spectra of (R)- and (S)-enantiomeric assemblies (0.1 mM) in the presence of Fe(II) (0.1 mM). (Reprinted with permission from [62], © 1999 American Chemical Society).

comprised of *bis*-terpyridine-di(ethylene glycol) and FeCl$_2$, the reversibility was investigated in a different way; decomplexation of the Fe(II)-polymer system could be shown by addition of the competitive ligand HEEDTA [67].

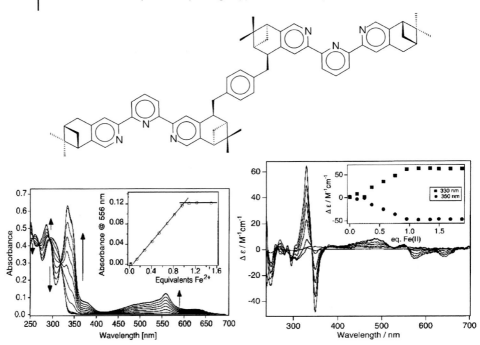

Figure 5.7 Top: chiral bridging ligand (–)-[ctpy-x-ctpy]; bottom: spectro-
photometric and CD-titration of (–)-[ctpy-x-ctpy] with Fe(BF$_4$)$_2$ (right).
(Reprinted with permission from [63], © 2001 American Chemical Society).

5.2.4
Polymers from Polymeric Building Blocks

In the previous sections, the construction of supramolecular polymers from well-
defined (monodisperse) organic building blocks was described; however, these
compounds were characterized by a high charge density, and, in general, their
mechanical properties were not particularly good. It has been a goal of many
researchers to combine the properties of known polymers with the features of
supramolecular entities. One route toward such materials is by polymeric
telechelics: terpyridine-functionalized polymers of fairly low molecular weight
can be used as building blocks for various polymeric superstructures such as
extended polymers or block and graft copolymers.

One of these approaches utilizes the facile nucleophilic aromatic substitution
of the readily available [68], but not cheap, 4′-chloroterpyridine, which can be
reacted very efficiently with different hydroxy- and thiol-terminated molecules in
the presence of KOH in DMSO to afford the corresponding oxo and thioethers in
high yields. As a first example an α-carboxy-ω-hydroxy-functionalized poly(oxytetramethylene) prepolymer was modified in this way with exactly one terpyridine
end unit (Scheme 5.16) [69, 70].

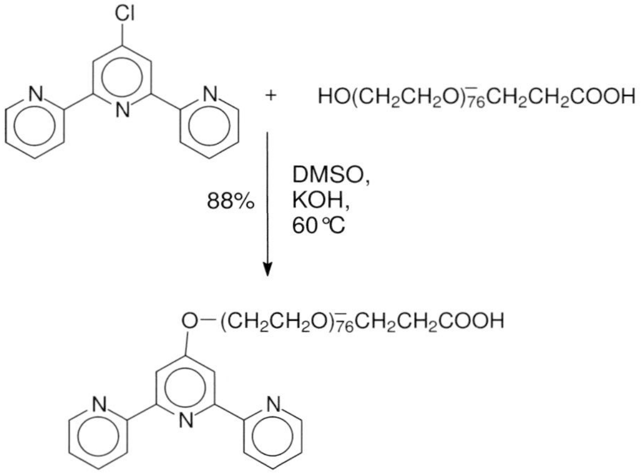

Scheme 5.16 Synthesis of an α,ω-functionalized metal complexing polymer.

The same strategy was applied to the modification of other different commercially available *bis*functionalized telechelics, based, e.g., on poly(oxytetramethylene) and poly(ethylene oxide) prepolymers (Scheme 5.17) [71–73].

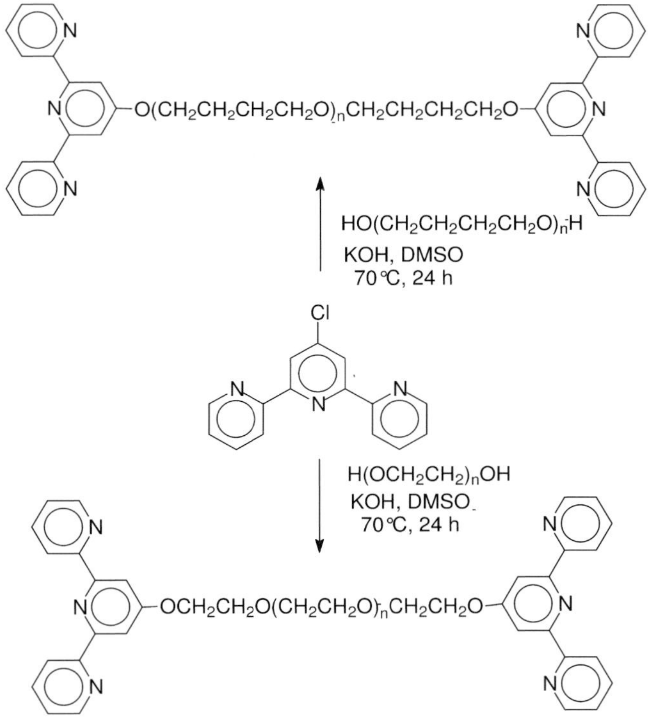

Scheme 5.17 Different metal-complexing compounds.

In all these cases, MALDI-TOF-MS was demonstrated to be a very powerful tool for investigating the resulting modified macromolecules. As shown in Figure 5.8, the complete *bis*functionalization of the poly(oxytetramethylene) with an average molecular weight $\bar{M}_n$ of 8000 Dalton can be easily proven by MALDI-TOF-MS. The difference in molecular weights between prepolymer (top) and the product (bottom) correlates with the complete reaction of two terpyridine units for each prepolymer; in addition, NMR, UV, and GPC results are in good agreement. This simple modification strategy can be easily expanded to other polymeric systems, such as poly(styrene), poly(methacrylate), or poly(siloxane)s.

The addition of octahedrally coordinating transition metal ions to terpyridine-modified prepolymers leads to a spontaneous self-assembly of two terpyridine units and, therefore, to a polymerization (in principle, a polyaddition reaction) (Scheme 5.18). Formation of these complexes can be reversed, e.g., by either changing pH [74] or applying electrochemical [75] or thermal changes [40]. The formation of terpyridine metal complexes and thus the non-covalent coordination

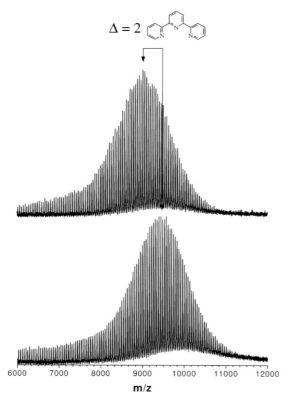

Figure 5.8 MALDI-TOF mass spectra of the α,ω-*bis*(hydroxy)-poly(ethylene oxide)$_{8000}$ and the product α,ω-*bis*(terpyridin-4'-yl)-poly(ethylene oxide)$_{8000}$ ($\bar{M}_n$ = 9375 g/mol); the difference in the peak maxima resembles two terpyridine units.

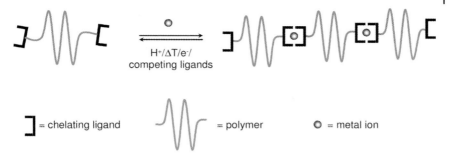

] = chelating ligand   = polymer   O = metal ion

Scheme 5.18 Schematic representation of the non-covalent polymerization.

polymers can easily be monitored by various techniques, such as UV/vis spectroscopy, NMR, or titration microcalorimetry.

Such a system, consisting of a high-molecular-weight poly(ethylene oxide) polymer end-capped with terpyridines, gave, upon addition of Fe(II) or Ni(II) acetate, an extended polymer, which was confirmed by the increase in viscosity (Figure 5.9) [76, 77]; the relatively weakly coordinating Cd(II) ions were studied analogously [78].

In order to investigate a variety of related polymers with different chain lengths, Fe(II)-containing polymers were synthesized for the first time using a combinatorial approach [76, 77]. Extended metallopolymers were also derived from Ru(II) ions. Solution viscosimetry suggested the appearance of high-molecular-weight polymers of ca. 130 000 Dalton (13 repeat units). A polyelectrolyte effect was present and could be eliminated by the addition of salt [79]. The material properties are significantly different compared to the powdery telechelic in that the extended supramolecular polymer possessed flexible, film-forming properties (Figure 5.10).

Another possibility for creating and investigating such systems is to first functionalize only one chain end with terpyridine and then to use either the directed or undirected coupling methods to obtain AA or AB and ABA block copolymer systems (Figure 5.11). Directed coupling can be achieved by first forming a *mono*-terpyridine metal complex; the most common metal for this strategy utilizes Ru(III). Subsequent reduction of Ru(III) to Ru(II) in the presence of a differently functionalized terpyridine leads to a heteroleptic complex. In contrast, undirected couplings use the same ligand for *bis*-complexation with bivalent metal salts. Looking at the AA homopolymer systems, this concept was recently realized using poly(ethylene oxide) functionalized with one terpyridine, which, upon complexation with various transition metal ions, gave water-soluble polymers with double the mass of the starting polymer ligand including the metal and counter ions [80]. These complexes were investigated with respect to their pH sensitivity, when it was found that the Fe(II)-, Co(II)-, Zn(II)- and Cd(II)-containing polymers showed decomplexation at high and low pH (13 and 1, respectively). Moreover, the Cu(II) polymer complex dissociated after the solution was kept at a low pH for a period of a few days; however, the metallo-homopolymers

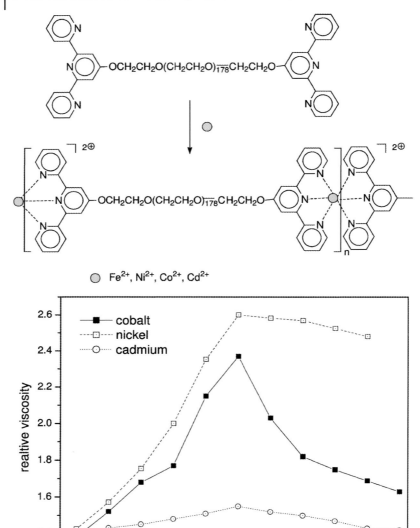

Figure 5.9 Top: synthesis of a metal-linked poly(ethylene glycol)$_{180}$.
Bottom: viscosity increase when adding Ni(II), Co(II), and Cd(II) acetate [77].

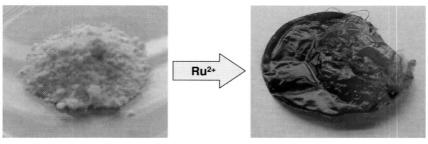

Figure 5.10 Photograph of a *bis*-terpyridine-functionalized poly(ethylene glycol) and the corresponding Ru(II) metallopolymer [79].

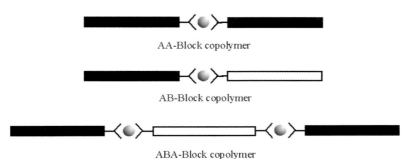

Figure 5.11 Schematic representation of AA-, AB- and ABA type metallo-supramolecular block copolymer systems.

containing Ru(II) and Ni(II) connectivity showed complete insensitivity to pH changes ranging from pH 0 to pH 14, even after several days. This showed that the viability of the $[M(tpy)_2]^{2+}$ linker was reversed by adjusting the pH in the case of selective metal ions. Thus, the existence of new avenues for possible applications of these systems as switchable "smart" materials should be noted. As for the characterization of these systems using mass spectrometry, the unfragmented *bis*-polymer complex as well as the fragmented free ligand could be detected only in the cases of Co(II) and Ru(II), which are among the most stable [81]. Increasing the laser intensity led to an increased ratio of free ligand/*bis*-polymer complex; this is not unexpected because of the partial dissociation of the complex, which is observed upon excitation.

Similar AA-type metallo-supramolecular block copolymers were recently synthesized using 4′-(hydroxymethyl)terpyridines, as initiators for the controlled coordinative ring-opening polymerization of lactides using aluminum alkoxide [61, 82, 83]. Since biodegradable polymers are becoming more important, especially in fields of tissue engineering and waste recycling, the development of new materials in that direction seems appropriate. Using this approach, polylactides end-capped with terpyridine were obtained with very narrow polydispersity. Complexation with Fe(II) led to the AA-metallopolymer, which could be characterized by means of GPC and MALDI-TOF-MS. Furthermore, as noted above, the stability of the Fe(II) complex is temperature dependent. Investigations concerning the temperature sensitivity of Fe(II)/polylactide complex films showed disappearance of the typical purple color at 160 °C; upon cooling the films, the color returns. This example also illustrates the reversibility of such metallo-supramolecular structures.

Concerning AB or ABA type structures, their preparation required directed coupling techniques. A well-known strategy for creating such hetero-complexes is again the Ru(III)/Ru(II) coupling method (see Chapter 3). First, the dark brown [Ru(4'-R-tpy)(Cl)$_3$] complex was formed by refluxing RuCl$_3$ in DMF with the 4'-functionalized terpyridine, and then this was treated with a different ligand under reductive conditions (EtOH/N-ethylmorpholine), resulting in the formation of the desired stable red Ru(II) hetero-complex. Applying this strategy to different terpyridine end-capped polymers led to the hetero-Ru(II)-complex polymers (Scheme 5.19) [84]. The synthesis of terpyridine-terminated chain-transfer agents produced specific macromolecular architectures by reversible-fragmentation chain-transfer (RAFT) polymerization; Ru-connectivity of the blocks generated supramolecular diblock macromolecules [85].

Both of the described metallo-supramolecular block copolymers showed a single peak when analyzed using GPC, indicating that no homopolymers were formed. Such AB type structures combining two different polymer chains have up to now only been accessible using living or controlled polymerization procedures. A library of AB-copolymers with varying lengths of the A and B blocks and the study of their morphology by AFM were recently reported (Figure 5.12) [86].

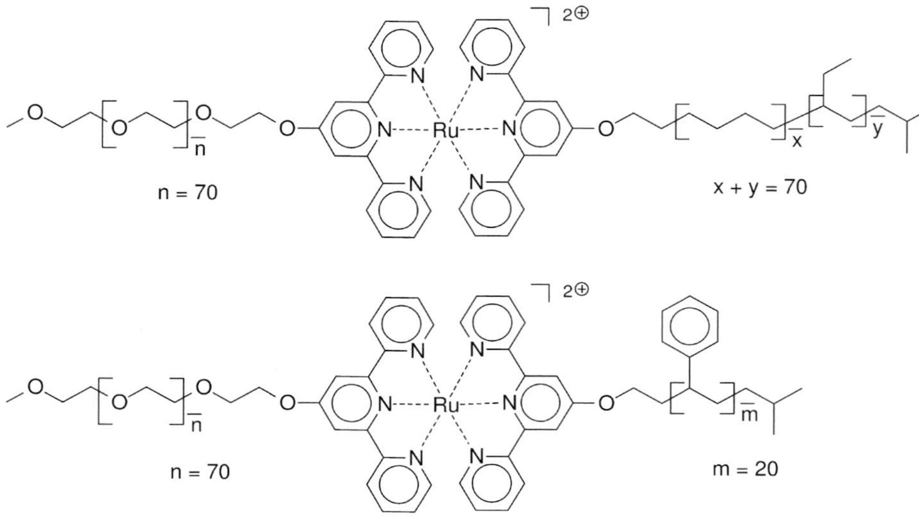

Scheme 5.19 Synthesized AB-type [Ru(tpy)(tpy')]$^{2+}$ complexes combining different polymer blocks [84].

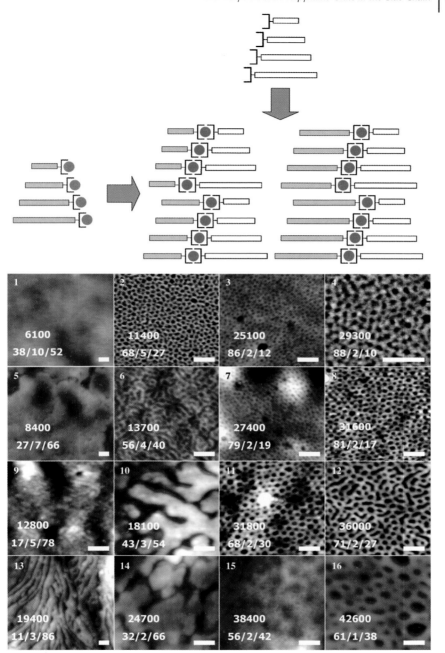

Figure 5.12 Schematic representation and AFM images of the block
copolymer library (upper number: molecular weight; lower numbers:
volume fraction block A/Ru-complex/block B; scale bar represents 100 nm).
(Reproduced with permission of The Royal Society of Chemistry).

The same strategy has also been applied for ABA type systems; thus, a *bis*-terpyridine end-capped poly(propylene oxide) telechelic was first complexed on both ends with RuCl$_3$ and subsequently complexed symmetrically on both sides with an appropriate 4′-functionalized terpyridine [68]. On the basis of MALDI-TOF-MS analysis, the existence of each species was shown by weight averages of the expected additional masses for each step of the functionalization. This route opens up new avenues for creating ABA systems, allowing for a wide range of combinations. It should be emphasized that the stability of these [Ru(tpy)(tpy′)]$^{2+}$-connected polymers is very high, so for many applications these complexes will not dissociate at the metal center.

In a recent publication, a powerful method for the preparation of *mono*- as well as *bis*-terpyridine-functionalized polymers has been presented in which a nitroxide-mediated controlled free-radical polymerization of styrene could be performed with a terpyridine-functionalized nitroxide. Well-defined polymers with a narrow polydispersity index were obtained [87]; moreover, the nitroxide-functionalized active end of the polymer could be quenched with a terpyridine-functionalized maleimide, resulting in a *bis*-terpyridine functionalized telechelic polymer (Scheme 5.20).

Two examples of metallopolymeric structures containing other metal ions should be mentioned. Perylene *bis*-imide dyes possessing two terpyridines gave rise to polymerization upon addition of Zn(II) [88]. Comparison of ^{1}H NMR spectra indicated the formation of polymeric structures, but since the stability of [Zn(tpy)$_2$]$^{2+}$ complexes is significantly lower than that of Ru(II) or Fe(II) [89], it might prove difficult to find hard evidence for the formation of long rod-like chains. Recently, Würthner et al. [90] prepared a series of related Zn(II) complexes and reported the results of UV/vis, ^{1}H NMR and isothermal titration calorimetry, and determined the ΔH values; the average polymer length was ascertained by AFM to be 15 repeat units, which correlated well with the NMR data suggesting >10 repeat units.

Recently, theoretical studies regarding [M(tpy)$_2$]$^{2+}$ polymers, including Monte Carlo simulations, were performed. The conclusion of the authors is that only at the stoichiometric point can high molecular weights be expected and that ring formation plays a minor role except at very low concentrations [91].

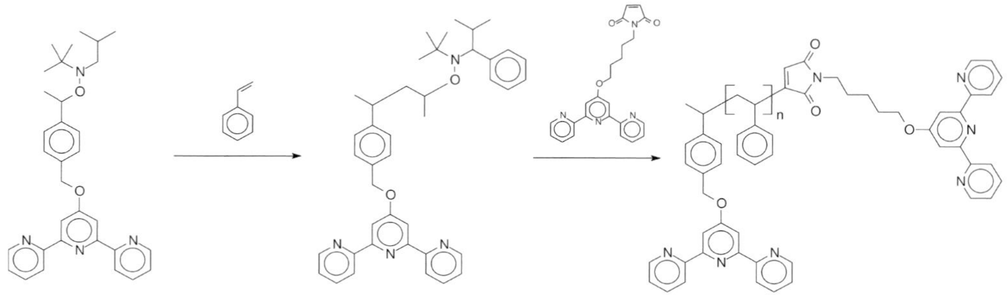

Scheme 5.20 Nitroxide-mediated controlled radical polymerization of styrene with a terpyridine-containing initiator and end-functionalization with terpyridine [87].

5.2.5
Other Systems

A different approach for generating polynuclear heterocyclic assemblies was discovered for a dinuclear platinum derivative, with the terpyridine acting as a new N,C∧C,N bridging ligand [92]. Although only observed in the crystal structure, the formation of infinite monodimensional organoplatinum chains through weak Pt(II)-Pt(II) contacts (3.283 Å) was demonstrated, with neighboring chains being kept together by graphitic (interplanar distance 3.52(2) Å) and van der Waals interactions (Figure 5.13). The knowledge about the existence of such chains in the crystal might provide helpful information for the utilization of these phenomena in the build-up of ordered chains or 2-dimensional crystals on different surfaces. Other related [Pt(R-tpy)(X)] complexes have appeared [93] and have been shown to possess interesting absorption and emission properties as well as assemblies.

In a new type of supramolecular polymer, $[M(tpy)_2]^{2+}$ complexes were combined with the quadruple H-bonding of ureidopyrimidinone (Figure 5.14) in the main chain. The properties of the metal complexes (e.g., photophysical or electrochemical properties) can be combined with the characteristics of the H-bonding moieties (e.g., reversibility and response to concentration, solvent, or temperature).

Ligands in which both supramolecular binding units were separated by a short spacer have been reported in a recent communication [94]. Zinc(II) as well as Fe(II) complexes were prepared and studied by UV-vis and NMR spectroscopy. The formation of polymeric species could be shown by a viscosimetric titration. The concept was extended to polymeric linkers in order to improve the properties of the polymer. Concentration-dependent viscosimetry revealed an exponential relationship of the viscosity with the concentration, a typical behavior for reversible supramolecular polymers; moreover, temperature-dependent rheometry showed high melt viscosities over a wide temperature range [95].

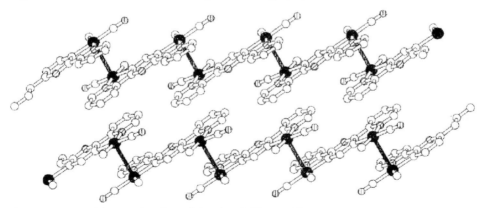

Figure 5.13 View of the crystal packing of dinuclear Pt(II) terpyridines showing the short Pt···Pt intermolecular interactions.
(Reprinted with permission from [92], © 2001 American Chemical Society).

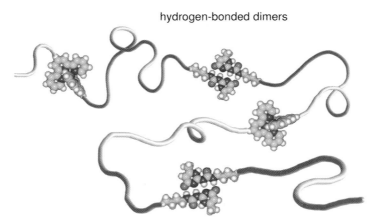

hydrogen-bonded dimers

Figure 5.14 Top: compounds containing terpyridine and ureidopyrimidinone [94]. Bottom: schematic representation of a supramolecular polymer containing both terpyridine metal complexes and quadruple *H*-bonding.

5.3
Biopolymers and Terpyridine Metal Complexes

Over the last two decades, there has been increasing interest in metal-chelating ligands, which can act as DNA/RNA intercalators or as inhibitors for certain enzymes, with the most prominent example being cisplatin [96, 97]. Terpyridines, as ligands, can also lead to potent intercalators, e.g., it was shown that for [Ru(tpy)(Cl)$_3$] adduct, the activity against L1210 leukemia cells is comparable with that of cisplatin [98]. Since there are numerous reviews on the issue of investigating different ligand systems, also including terpyridines for such purposes, the focus will be only on covalently bonded systems. As far as the investigation of enzyme-polymer hybrid systems is concerned, recent approaches made include metal-to-ligand complexation using terpyridine ligands.

It has been shown that Ru(II) complexes of the type [Ru(tpy)(dppz)(MeCN)]$^{2+}$ (dppz = dipyrido[3,2-*a*:2′,3′-*c*]phenazine), which are tethered to an oligonucleotide (ODN) strand (Scheme 5.21), can be specifically photolyzed to give a reactive aqua derivative, which can then form a duplex with a DNA target 11mer [99]. This duplex was found to be significantly more stable than the natural, non-Ru(II) complex containing a DNA • DNA duplex.

Following the hybridization of the ODN with DNA, the nitrogen of a guanine completes the octahedral coordination sphere of the mixed Ru(II) complex;

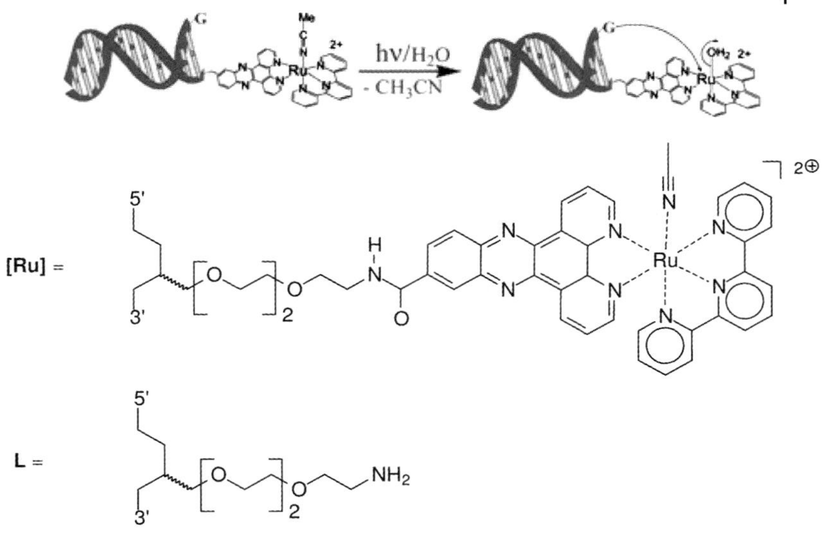

[Ru] =

L =

oligo			yield (%)	
			solid-phase synthesis	postsynthetic labeling
antisense ODNs	natural	5'-CTTACCAATC-3'	67	
	5'-modified	5'-L-pCTTACCAATC-3'	56	
		5'-Ru-pCTTACCAATC- 3'		46
	3'-modified	5'-CTTACCAATCp-L-3'	59	
		5'-CTTACCAATCp-[Ru]- 3'	69	51
	middle	5'-CTTACp-L-pCAATC- 3'	87	
		5'-CTTACp-Ru-pCAATC-3'		28
	5'-modified	5'-L-pCTTACCAATCp- **L**-3'	64	
		5'-[Ru]-pCTTACCAATCp -Ru-3'		34
target 11mer		5'-TGATTGGTAAG-3'	87	

Scheme 5.21 Yields of the synthetic [Ru(tpy)$_2$]$^{2+}$-labeled oligonucleotide conjugates and their target oligo-DNA [99].

however, because of the steric hindrance, these complexes are vulnerable to ligand exchange with, for example, pyridine. Therefore, the reactivity and specificity of the resulting aqua-Ru(II)-ODN conjugates can be controlled.

Another example of the useful incorporation of a terpyridine complex into an oligonucleotide was shown by Bashkin et al. A 17-mer oligonucleotide probe containing a terpyridine attached to a serinol, which can act as a building block in DNA sequencing, was designed in order to target a 159-mer fragment of the HIV *gag* gene messenger RNA (Scheme 5.22) [100].

Experiments showed that, upon complexation of the terpyridine probe with Cu(II), the target mRNA was specifically cleaved after forming a duplex. Several different 17-mer DNA probe sequences were prepared via automated DNA synthesis with the serinol-terpyridine being included at a different position for

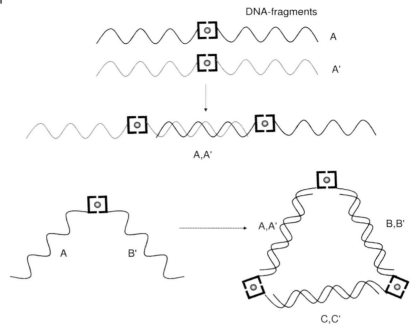

Scheme 5.24 Polymers and triangular arrays of DNA fragments and terpyridine complexes [102, 103].

Another application of metal complexes containing terpyridine as ligand is the determination of electron transfer (ET) rates in Azurin from *Pseudomonas aeruginosa*. This protein is widely studied as a model electron-transfer protein, in particular with respect to the coordination of the copper ion [Cu(II) ↔ Cu(I)]. Gray et al. studied ET transfer rates of Ru(II)-modified azurins [Ru(tpy)(bpy/phen)(His83)Az]$^{2+}$ [105] and showed that the time constants for electron tunneling [Cu(I) → Ru(III)] in crystals were roughly the same as those measured in solution, indicating very similar protein structures in the two states.

As a means of combining catalytic activity of enzymes with metallo-supra-molecular ordering, the first experiments with respect to functionalizing an enzyme and a protein with a terpyridine unit have been conducted and, in the case of the enzyme, tested for catalytic activity [106]. A maleimide-functionalized terpyridine was reacted with free thiol groups from lipase B from *Candida antarctica* or the free thiol groups of Bovine Serum Albumin, which were obtained after reduction of exposed disulfide bonds. The functionalized biomolecules were then coupled through *bis*terpyridine metal [Fe(II), Ru(II)] complexation, which was followed by means of UV/Vis-spectroscopy. This approach represents a first step toward the ability to order and aggregate such biomolecules in a specific manner. Future experiments, such as the coupling of biomolecules to polymers, should open up new possibilities for the development of bio-supramolecular structures.

Recent work from Schubert et al. was directed toward sensor applications utilizing $[M(tpy)(tpy')]^{2+}$ complexation in combination with the well-known biotin-avidin system [107, 108]. A terpyridine was functionalized with a biotin (vitamin H) unit in the 4'-position with a short pentyl spacer as well as a long PEG_{75} spacer, opening up possibilities of functionalization of surfaces with an activated layer of avidin.

References

1 L. Brunsveld, B. J. B. Folmer, E. W. Meijer, *MRS Bulletin* **2000**, *25*, 49–53.

2 U. S. Schubert, C. Eschbaumer, *Angew. Chem. Int. Ed.* **2002**, *41*, 2892–2926.

3 M. Rehahn, *Acta Polym.* **1998**, *49*, 201–224.

4 G. R. Newkome, A. K. Patri, E. Holder, U. S. Schubert, *Eur. J. Org. Chem.* **2004**, 235–254.

5 H. Nishihara, T. Shimura, A. Ohkubo, N. Matsuda, K. Aramaki, *Adv. Mater.* **1993**, *5*, 752–754.

6 P. Nguyen, P. Gomez-Elipe, I. Manners, *Chem. Rev.* **1999**, *99*, 1515–1548.

7 R. P. Kingsborough, T. M. Swager, *Prog. Inorg. Chem* **1999**, *48*, 123–231.

8 H. Frauenrath, *Prog. Polym. Sci.* **2005**, *30*, 325–384.

9 H. Hofmeier, U. S. Schubert, *Chem. Commun.* **2005**, 2423–2432.

10 D. R. McMillin, J. J. Moore, *Coord. Chem. Rev.* **2002**, *229*, 113–121.

11 K. E. Erkkila, D. T. Odom, J. K. Barton, *Chem. Rev.* **1999**, *99*, 2777–2795.

12 K. T. Potts, D. A. Usifer, *Macromolecules* **1988**, *21*, 1985–1991.

13 K. T. Potts, K. A. Usifer, A. Guadelupe, H. D. Abruña, *J. Am. Chem. Soc.* **1987**, *109*, 3961–3967.

14 K. Hanabusa, K. Nakano, T. Koyama, H. Shirai, N. Hojo, A. Kurose, *Makromol. Chem.* **1990**, *191*, 391–396.

15 K. Hanabusa, A. Nakamura, T. Koyama, H. Shirai, *Makromol. Chem.* **1992**, *193*, 1309–1319.

16 K. J. Calzia, G. N. Tew, *Macromolecules* **2002**, *35*, 6090–6093.

17 L. R. Rieth, R. F. Eaton, G. W. Coates, *Angew. Chem. Int. Ed.* **2001**, *40*, 2153–2156.

18 M. Heller, U. S. Schubert, *Macromol. Rapid Commun.* **2002**, *23*, 411–415.

19 U. S. Schubert, H. Hofmeier, *Macromol. Rapid Commun.* **2002**, *23*, 561–566.

20 H. Hofmeier, U. S. Schubert, *Macromol. Chem. Phys.* **2003**, *204*, 1391–1397.

21 H. Hofmeier, A. El-Ghayoury, U. S. Schubert, *e-Polymers* **2003**, *53*, 1–15.

22 M. A. R. Meier, U. S. Schubert, *J. Polym. Sci. Part A., Polym. Chem.* **2003**, *41*, 2964–2973.

23 Y.-S. Cho, H.-K. Lee, J.-S. Lee, *Macromol. Chem. Phys.* **2002**, *203*, 2495–2500.

24 A. El-Ghayoury, H. Hofmeier, B. de Ruiter, U. S. Schubert, *Macromolecules* **2003**, *36*, 3955–3959.

25 H. Hofmeier, A. El-Ghayoury, U. S. Schubert, *J. Polym. Sci. Part A., Polym. Chem.* **2004**, *42*, 4028–4035.

26 M. Kimura, T. Horai, K. Hanabusa, H. Shirai, *Adv. Mater.* **1998**, *10*, 459–462.

27 C. T. Wong, W. K. Chan, *Adv. Mater.* **1999**, *11*, 455–459.

28 U. S. Schubert, C. Eschbaumer, C. H. Weidl, *Design. Monom. Polym.* **1999**, 2, 185–189.

29 U. S. Schubert, M. Heller, *Chem. Eur. J.* **2001**, 7, 5252–5259.

30 B. G. G. Lohmeijer, U. S. Schubert, *J. Polym. Sci., Part A: Polym. Chem.* **2003**, 41, 1413–1427.

31 E. C. Constable, *Macromol. Symp.* **1995**, 8, 503–524.

32 P. F. H. Schwab, M. D. Levin, J. Michl, *Chem. Rev.* **1999**, 99, 1863–1933.

33 E. C. Constable, A. M. W. Cargill Thompson, D. A. Tochter, *Supramolecular Chemistry*, Kluwer Academic Press, Dordrecht, **1992**.

34 M. Schütte, D. G. Kurth, M. R. Linford, H. Cölfen, H. Möhwald, *Angew. Chem. Int. Ed.* **1998**, 37, 2891–2893.

35 C. Tziatzios, H. Durchschlag, C. H. Weidl, C. Eschbaumer, W. Mächtle, P. Schuck, U. S. Schubert, D. Schubert, *Ultracentrifugation Studies on the Solution Properties of Supramolecular Building Blocks for Polymers: Potential, Problems and Solutions*, in *Synthetic Macromolecules with Higher Structural Order* (Ed.: I. Kahn), **2001**, pp. 185–200.

36 D. G. Kurth, R. Osterhout, *Langmuir* **1999**, 15, 4842–4846.

37 T. Salditt, Q. An, A. Plech, C. Eschbaumer, U. S. Schubert, *Chem. Commun.* **1998**, 2731–2732.

38 T. Salditt, Q. An, A. Plech, J. Peisl, C. Eschbaumer, C. H. Weidl, U. S. Schubert, *Thin Solid Films* **1999**, 354, 208–214.

39 U. S. Schubert, C. Eschbaumer, Q. An, T. Salditt, *Polym. Prepr.* **1999**, 40, 414–415.

40 U. S. Schubert, C. Eschbaumer, Q. An, T. Salditt, *J. Incl. Phenom.* **1999**, 32, 35–43.

41 U. S. Schubert, C. Eschbaumer, C. H. Weidl, A. B. Vix, T. Salditt, *PMSE (ACS), Proc.* **2001**, 84, 455–456.

42 D. G. Kurth, P. Lehmann, M. Schütte, *Proc. Natl. Acad. Sci.* **2000**, 97, 5704–5707.

43 H. Krass, E. A. Plummer, J. M. Haider, P. R. Barker, N. W. Alcock, Z. Pikramenou, M. J. Hannon, D. G. Kurth, *Angew. Chem. Int. Ed.* **2001**, 40, 3862–3865.

44 D. G. Kurth, J. P. López, W.-F. Dong, *Chem. Commun.* **2005**, 2119–2121.

45 M. Schütte, D. G. Kurth, M. R. Linford, H. Cölfen, H. Möhwald, *Angew. Chem. Int. Ed.* **1998**, 37, 2891–2893.

46 D. G. Kurth, M. Schütte, J. Wen, *Colloids Surf., A* **2002**, 198–200, 633–643.

47 Y. Bodenthin, U. Pietsch, H. Möhwald, D. G. Kurth, *J. Am. Chem. Soc.* **2005**, 127, 2110–3114.

48 T. E. Janini, J. L. Fattore, D. L. Mohler, *J. Organomet. Chem.* **1999**, 578, 260–263.

49 W. Y. Ng, W. K. Chan, *Adv. Mater.* **1997**, 9, 716–719.

50 W. Y. Ng, X. Gong, W. K. Chan, *Chem. Mater.* **1999**, 11, 1165–1170.

51 W. K. Chan, X. Gong, W. Y. Ng, *Appl. Phys. Lett.* **1997**, 71, 2919–2921.

52 P. K. Ng, W. Y. Ng, X. Gong, W. K. Chan, *Mater. Res. Soc. Symp. Proc.* **1998**, 488, 581–586.

53 S. Kelch, M. Rehahn, *Macromolecules* **1999**, *32*, 5818–5828.

54 S. Kelch, M. Rehahn, *Chem. Commun.* **1999**, 1123–1124.

55 O. Schmelz, M. Rehahn, *e-Polymers* **2002**, *47*, 1–29.

56 A. El-Ghayoury, A. P. H. J. Schenning, E. W. Meijer, *J. Polym. Sci., Part A: Polym. Chem.* **2002**, *40*, 4020–4023.

57 J. A. Barron, S. Glazier, S. Bernhard, K. Takada, P. L. Houston, H. D. Abruña, *Inorg. Chem.* **2003**, *42*, 1448–1455.

58 J. Hjelm, E. C. Constable, E. Figgemeier, A. Hagfeldt, R. Handel, C. E. Housecroft, E. Mukhtar, E. Schofield, *Chem. Commun.* **2002**, 284–285.

59 G. D. Storrier, S. B. Colbran, *J. Chem. Soc., Dalton Trans.* **1996**, 2185–2186.

60 G. D. Storrier, S. B. Colbran, D. C. Craig, *J. Chem. Soc., Dalton Trans.* **1997**, 3011–3028.

61 M. Heller, U. S. Schubert, *Macromol. Rapid Commun.* **2001**, *22*, 1358–1363.

62 M. Kimura, M. Sano, T. Muto, K. Hanabusa, H. Shirai, N. Kobayashi, *Macromolecules* **1999**, *32*, 7951–7953.

63 S. Bernhard, K. Takada, D. J. Diaz, H. D. Abruña, H. Murner, *J. Am. Chem. Soc.* **2001**, *123*, 10265–10271.

64 E. C. Constable, C. E. Housecroft, C. B. Smith, *Inorg. Chem. Commun.* **2003**, *6*, 1011–1013.

65 P. R. Andres, U. S. Schubert, *Synthesis* **2004**, 1229–1238.

66 P. R. Andres, U. S. Schubert, *Macromol. Rapid Commun.* **2004**, *25*, 1371–1375.

67 S. Schmatloch, M. F. Gonzalez, U. S. Schubert, *Macromol. Rapid Commun.* **2002**, *23*, 957–961.

68 U. S. Schubert, S. Schmatloch, A. A. Precup, *Des. Monomers Polym.* **2002**, *5*, 211–221.

69 U. S. Schubert, C. Eschbaumer, *Polym. Prepr.* **1999**, *40*, 1070–1071.

70 U. S. Schubert, C. Eschbaumer, *Macromol. Symp.* **2001**, *163*, 177–187.

71 U. S. Schubert, C. Eschbaumer, *Polym. Prepr.* **2000**, *41*, 542–543.

72 U. S. Schubert, O. Hien, C. Eschbaumer, *Macromol. Rapid Commun.* **2000**, *21*, 1156–1161.

73 C. H. Weidl, A. A. Precup, C. Eschbaumer, U. S. Schubert, *PMSE (ACS), Proc.* **2001**, *84*, 649–650.

74 R. Farina, R. Hogg, R. G. Wilkins, *Inorg. Chem.* **1968**, *7*, 170–172.

75 G. Hochwimmer, Ph.D. thesis, Technische Universität München (Munich), **1999**.

76 U. S. Schubert, S. Schmatloch, *Polym. Prepr.* **2001**, *42*, 395–396.

77 S. Schmatloch, A. M. J. van den Berg, A. S. Alexeev, H. Hofmeier, U. S. Schubert, *Macromolecules* **2003**, *36*, 9943–9949.

78 J. van der Gucht, N. A. M. Besseling, H. P. van Leeuwen, *J. Phys. Chem. B* **2004**, *108*, 2531–2539.

79 H. Hofmeier, S. Schmatloch, D. Wouters, U. S. Schubert, *Macromol. Chem. Phys.* **2003**, *204*, 2197–2203.

80 B. G. G. Lohmeijer, U. S. Schubert, *Macromol. Chem. Phys.* **2003**, *204*, 1072–1078.

81 M. A. R. Meier, B. G. G. Lohmeijer, U. S. Schubert, *J. Mass Spectrom.* **2003**, *38*, 510–516.

82 M. Heller, U. S. Schubert, *e-Polymers* **2002**, *27*, 1–11.

83 M. Heller, U. S. Schubert, *Macromol. Symp.* **2002**, *177*, 87–96.

84 B. G. G. Lohmeijer, U. S. Schubert, *Angew. Chem. Int. Ed.* **2002**, *41*, 3825–3829.

85 G. Zhou, I. I. Harruna, *Macromolecules* **2005**, *38*, 4114–4123.

86 B. G. G. Lohmeijer, U. S. Schubert, *Chem. Commun.* **2004**, 2886–2887.

87 B. G. G. Lohmeijer, U. S. Schubert, *J. Polym. Sci. Part A., Polym. Chem.* **2004**, *42*, 4016–4024.

88 R. Dobrawa, F. Würthner, *Chem. Commun.* **2002**, 1878–1879.

89 R. H. Holyer, C. D. Hubbard, S. F. A. Kettle, R. G. Wilkins, *Inorg. Chem.* **1966**, *5*, 622–625.

90 R. Dobrawa, M. Lysetska, P. Ballester, M. Grüne, F. Würthner, *Macromolecules* **2005**, *38*, 1315–1325.

91 C.-C. Chen, E. E. Dormidontova, *J. Am. Chem. Soc.* **2004**, *126*, 14972–14978.

92 A. Doppiu, G. Minghetti, M. A. Cinellu, S. Stoccoro, A. Zucca, M. Manassero, *Organometallics* **2001**, *20*, 1148–1152.

93 X.-J. Liu, J.-K. Feng, J. Meng, Q.-J. Pan, A.-M. Ren, X. Zhou, H. X. Zhang, *Eur. J. Inorg. Chem.* **2005**, 1856–1866.

94 H. Hofmeier, A. El-Ghayoury, A. P. H. J. Schenning, U. S. Schubert, *Chem. Commun.* **2004**, 318–319.

95 H. Hofmeier, R. Hoogenboom, M. E. L. Wouters, U. S. Schubert, *J. Am. Chem. Soc.* **2005**, *127*, 2913–2921.

96 E. Reed, *Cancer Chemoth. Biol. Response Modif.* **1999**, *18*, 144–151.

97 E. E. Trimmer, J. M. Essigmann, *Essays Biochem.* **1999**, *34*, 191–211.

98 P. M. van Vliet, S. M. S. Toekimin, J. G. Haasnoot, J. Reedijk, O. Novakova, O. Vrana, V. Brabec, *Inorg. Chim. Acta* **1995**, *231*, 57–64.

99 D. Ossipov, S. Gohil, J. Chattopadhyaya, *J. Am. Chem. Soc.* **2002**, *124*, 13416–13433.

100 A. T. Daniher, J. K. Bashkin, *Chem. Commun.* **1998**, 1077–1078.

101 H. Hakala, E. Maeki, H. Loennberg, *Bioconjugate Chem.* **1998**, *9*, 316–321.

102 K. M. Stewart, L. W. McLaughlin, *Chem. Commun.* **2003**, 2934–2935.

103 J. S. Choi, C. W. Kang, K. Jung, J. W. Yang, Y.-G. Kim, H. Han, *J. Am. Chem. Soc.* **2004**, *126*, 8606–8607.

104 G. Bianké, R. Häner, *ChemBioChem* **2004**, *5*, 1063–1068.

105 B. R. Crane, A. J. Di Bilio, J. R. Winkler, H. B. Gray, *J. Am. Chem. Soc.* **2001**, *123*, 11623–11631.

106 K. Velonia, P. Thordarson, P. R. Andres, U. S. Schubert, A. E. Rowan, R. J. M. Nolte, *Polym. Prepr.* **2003**, *44*, 648.

107 H. Hofmeier, J. Pahnke, C. H. Weidl, U. S. Schubert, *Biomacromolecules* **2004**, *5*, 2055–2064.

108 S. Schmatloch, C. H. Weidl, I. van Baal, J. Pahnke, U. S. Schubert, *Polym. Prepr.* **2002**, *43*, 684–685.

6

Functional 3-D Architectures Based on Terpyridine Complexes

6.1
Introduction

In the quest for new functional materials, supramolecular metallodendrimers, micelles, and resins have played an ever-increasing role. Dendrimers [1–3] represent a special class of macromolecules, which possess low polydispersity (ideally monodisperse), nearly spherical topology, internal void regions for host-guest chemistry, and unique tailorable surface characteristics. In particular, metal-containing dendrimers are of intense current interest because of their potential use as catalysts [4–9] or molecular carriers for catalysts and light-harvesting arrays [10–14]. Generally, metallodendrimers can be divided into two general categories: those having the metal encapsulated within the dendritic superstructure and those having the metal located at or near the dendrimer surface. Since this monograph is specifically directed toward terpyridine-metal-terpyridine connectivity, information regarding metallodendrimers [15] with other than terpyridine connecting units should be sought from other numerous reviews and monographs [7, 9–11, 16–25].

   Whereas structurally perfect dendrimers are, in essence, precise unimolecular micelles [26, 27], micelles, on the other hand, can be envisioned as an assembly of polymers, and it is this superstructure that can be used for similar molecular encapsulation applications [28–30]. Since micelles are reversible assemblies (in contrast to dendrimers), they can be easily disassembled under diverse conditions. Such aggregates have been utilized for innumerable applications, from detergents to drug delivery. Polymeric micelles are in between hyperbranched (imperfect) dendritic-like (branched) materials and micelles, which leads finally to functional resins; all of these assemblies find applications in heterogeneous catalysis.

   In the first section of this chapter, dendrimers bearing terpyridine units are discussed, and specific nanostructures are considered. In 1993, the first dendrimer was assembled via *bis*terpyridine-Ru(II) "$[Ru(tpy)_2]^{2+}$" connectivity when Newkome et al. [31] utilized Constable's 4′-chloroterpyridine [32] in the construction process. The formation of appropriate dendrons with crucially located terpyridine moieties was devised. Scheme 6.1 shows the procedure that generated the initial metallo-dendritic superstructure via the metal-centered assembly process. This con-

Modern Terpyridine Chemistry. U. S. Schubert, H. Hofmeier, G. R. Newkome
Copyright © 2006 WILEY-VCH Verlag GmbH & Co. KGaA, Weinheim
ISBN: 3-527-31475-X

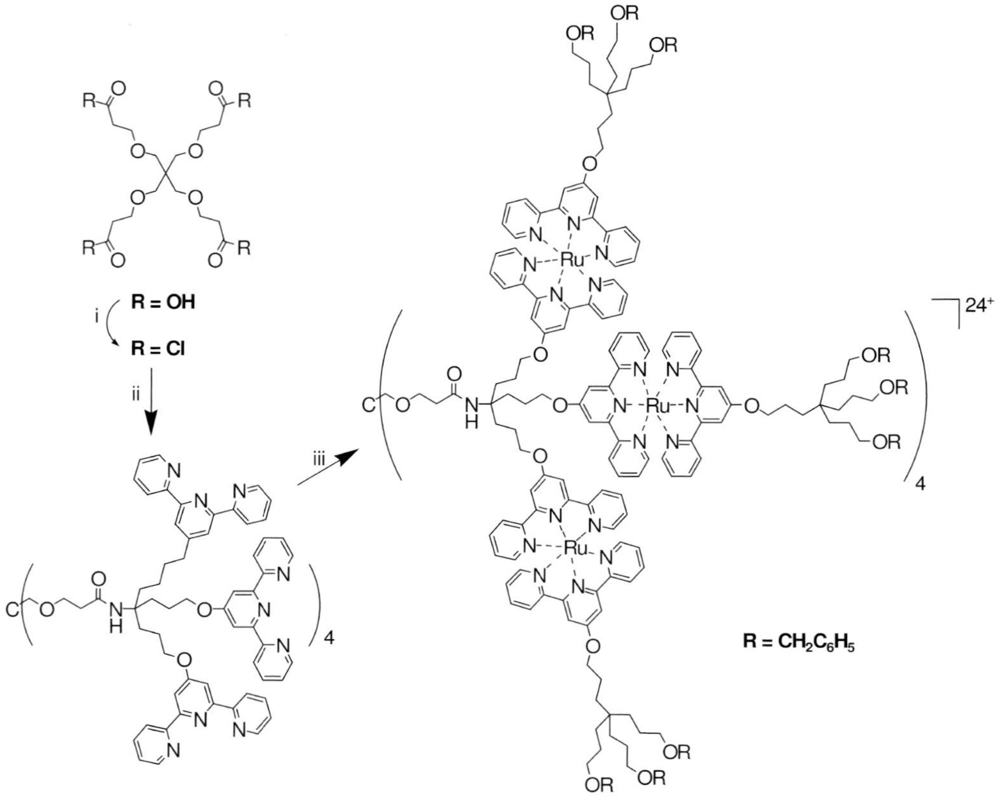

Scheme 6.1 (i) SOCl$_2$, CH$_2$Cl$_2$, 45 °C, 13 h; (ii) Et$_3$N, *tris*-tpy dendron (4 equiv.), 25 °C, 3 days; (iii) N-ethylmorpholine, [Ru(4′-R-tpy)(Cl)$_3$] dendron (15 equiv.), MeOH, reflux, 1.5 h.

struction technique, as noted in the previous chapters, permits precise step-wise assembly, since each (macromolecular) component (core or dendron) can be full characterized; then each [Ru(tpy)$_2$]$^{2+}$ connection can be quantitatively ascertained by NMR, in which the free ligand and complex possess distinctive chemical shifts. This simple assembly process permits access to precise macromolecules, which expands the resultant materials into the micron-level regime. Interestingly, the presence of a single diamagnetic metal center accurately predicts the degree of formation of the predicted superstructure.

The second part of this chapter describes polymeric micelles that are composed of terpyridine-containing polymers in which less-than-perfect materials relative to a dendrimer's perfection are generated. By preparing amphiphilic block and graft copolymers through terpyridine complexation, micelles can be obtained. And in the last part, resins (e.g., polymeric microbeads) modified with terpyridine units are discussed as well as the subsequent complexation, which encompasses the micron scale.

6.2
Dendrimers Containing Terpyridine Metal Complexes in the Dendrimer Core

Newkome et al. [33–35] utilized the incorporation of tpy-Ru(II)-tpy moiety to connect two independently prepared and different dendrons, mimicking a lock and key system; this was presented at the 1995 Centennial Anniversary Symposium of Emil Fischer's original "lock and key" concept in Birmingham, England. In this example, the construction of the "key" used a long alkyl spacer between the [Ru(4'-R-tpy)(Cl)$_3$] monocomplex and the 2nd generation dendron, and the "lock" used a 3rd generation dendron bearing an internal free terpyridine ligand attached to the central focal point. Thus, only a key possessing the Ru(III) site, which is capable of in situ reduction to the "terpyridine-Ru(II)" locus, can be connected to the internal lock (Scheme 6.2). A series of locks and keys permitted the evaluation of the meaning of "inside a dendrimer"; the electrochemistry demonstrated the internal (isolated) location of the [Ru(tpy)$_2$]$^{2+}$ site; the coupling of the lower generation components, in which this connection is "outside the dendrimer" exhibited an obvious reversible cyclic voltammogram.

Also in 1995, Chow et al. (Figure 6.1) used [Fe(tpy)$_2$]$^{2+}$ connectivity to create the central metal core for a dendrimer based on the convergent mode of construction. They reported the synthesis and characterization of metallodendrimers using benzyl ether-based (Fréchet-type) dendritic building blocks possessing propylene spacer moieties up to the 4th generation with an [Fe(tpy)$_2$]$^{2+}$ central core [36–39].

Recently, Chow et al. [40] described a series of homo- and heteroleptic benzyl ether dendrimers with an [Ru(tpy)$_2$]$^{2+}$ connectivity, which is central for the assembly of symmetrical homoleptic complexes as well as related unsymmetrical heteroleptic complexes (Figure 6.2). Cyclic voltammetry (CV) measurements were conducted in order to investigate the influence of the polyethereal dendritic fragments on the redox potentials of the [Ru(tpy)$_2$]$^{2+}$ core as well as to investigate the influence of the shape of the dendrimer on redox reversibility. No induction effects were observed on the redox potentials of the electrochemically active unit which could have arisen from the electron-rich polyethers. Also no preferred orientation of the non-spherical different dendrimers towards the electrode could be observed. The redox reversibility decrease was correlated with the size exclusion chromatographic data of this metallodendrimer family.

Constable et al. [41] reported the synthesis of [Ru(tpy)$_2$]$^{2+}$ dendrimers of the "Fréchet type". First, second, and third generation dendrons were coupled to give the corresponding [Co(tpy)$_2$]$^{2+}$ and [Fe(tpy)$_2$]$^{2+}$ complexes, which were also characterized. Investigation of these structures possessing the metallo-core was performed by X-ray analysis for the first-generation species; molecular modeling studies were performed on the higher-generation species (Figure 6.3).

Kimura et al. [42] reported the use of [Ru(tpy)$_2$]$^{2+}$ connectivity to couple two 1,3,5-phenylene-based dendrons; the resultant dendrimer exhibited one oxidation ($E_{1/2}$ = +1.12 V vs SCE) and two reduction processes ($E_{1/2}$ = −1.28 and −1.43 V vs SCE). A large voltage difference between the current maximum of the reduction

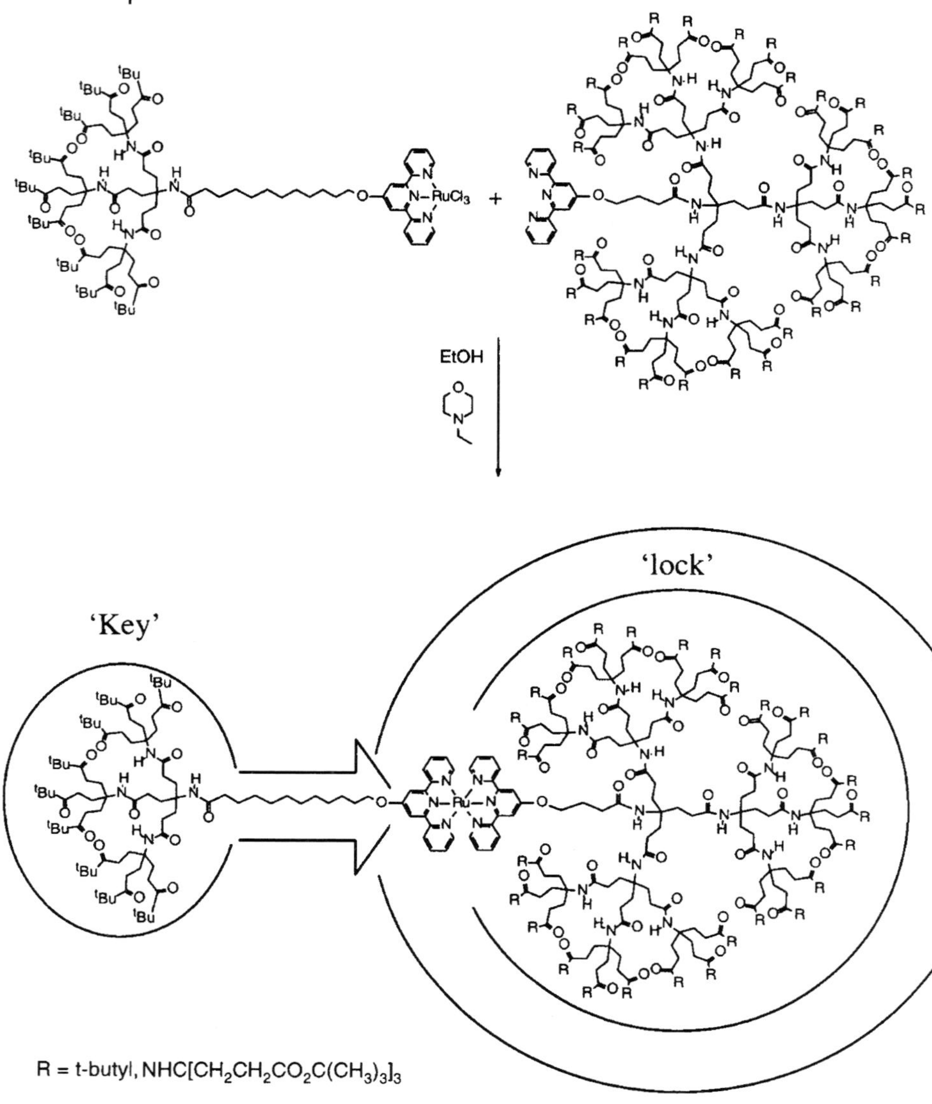

EtOH

Scheme 6.2 A key and lock system utilizing [Ru(tpy)$_2$]$^{2+}$ connectivity.
(Reprinted with permission from [33, 34]).

and oxidation wave (ΔE) indicated a slower electron transfer compared to the non-dendritic complex.

Diederich et al. [43] have recently prepared a series of homo- and heteroleptic [Ru(tpy)$_2$]$^{2+}$ complexes possessing hydrophilic and hydrophobic dendrons with the goal of developing amphiphilic vectors for potential gene delivery. Because of the step-wise mode of construction (Figure 6.4), dendrons can possess the same or different degrees of lipophilic or hydrophilic character.

Figure 6.1 Chow's benzyl ether-based dendrimer with an $[Fe(tpy)_2]^{2+}$ core [36–39].

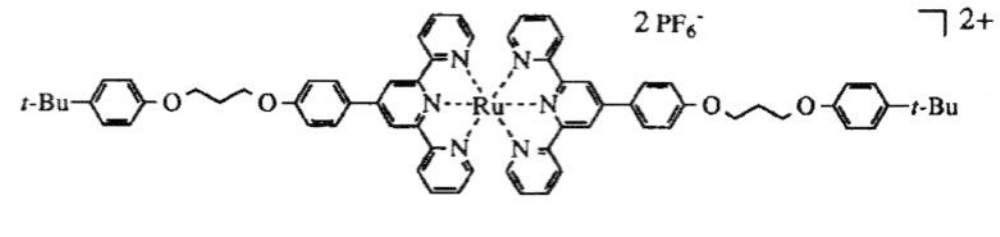

$[Ru(G0)_2]^{2+}(PF_6)^-_2$ **1**

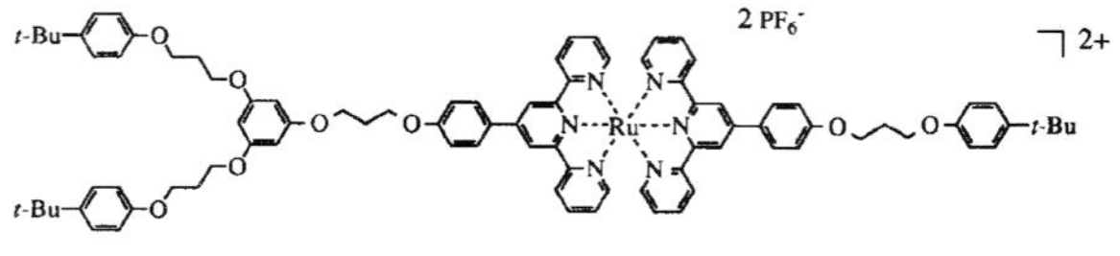

$[Ru(G0)(G1)_2]^{2+}(PF_6)^-_2$ **2**

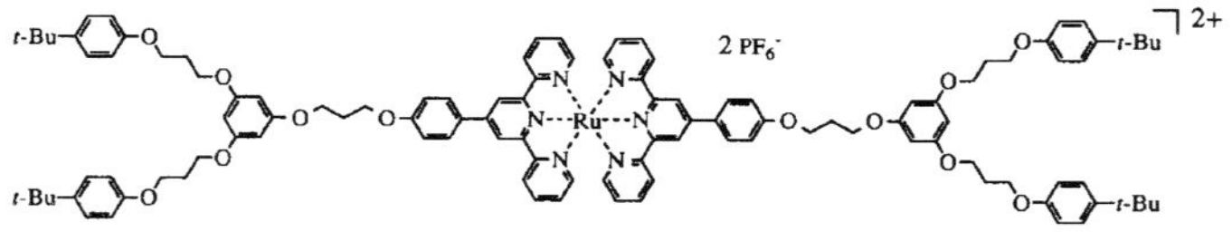

$[Ru(G1)_2]^{2+}(PF_6)^-_2$ **3**

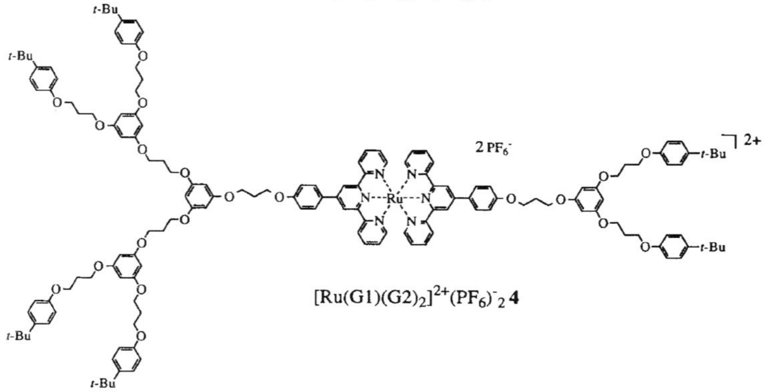

$[Ru(G1)(G2)_2]^{2+}(PF_6)^-_2$ **4**

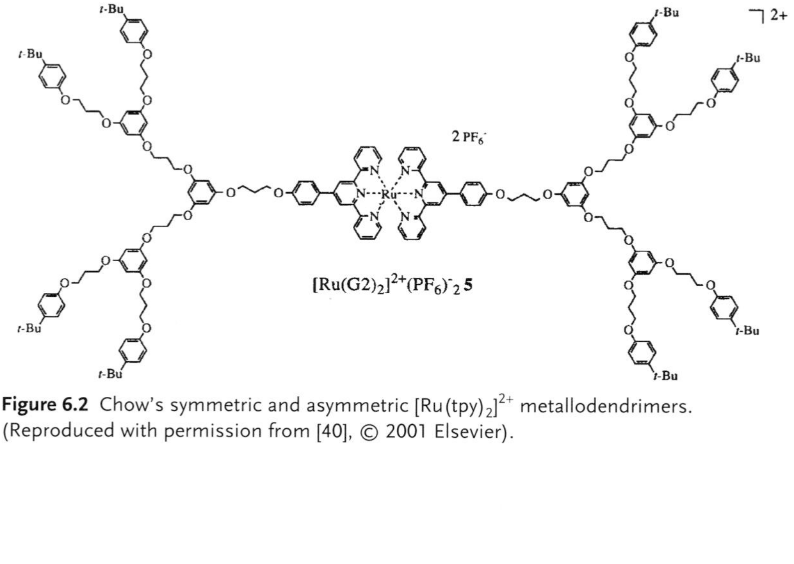

$[Ru(G2)_2]^{2+}(PF_6)^-_2$ **5**

Figure 6.2 Chow's symmetric and asymmetric $[Ru(tpy)_2]^{2+}$ metallodendrimers. (Reproduced with permission from [40], © 2001 Elsevier).

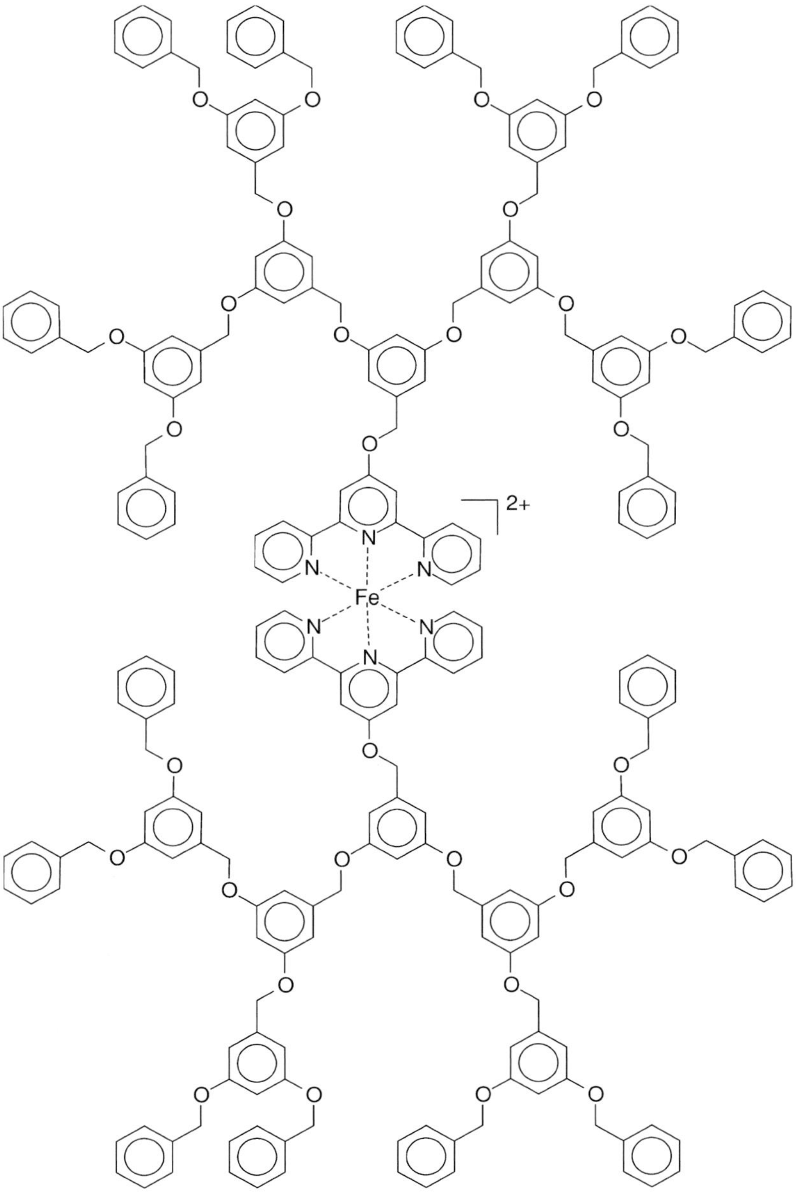

Figure 6.3 Third-generation dendrimer by Constable et al. [41].

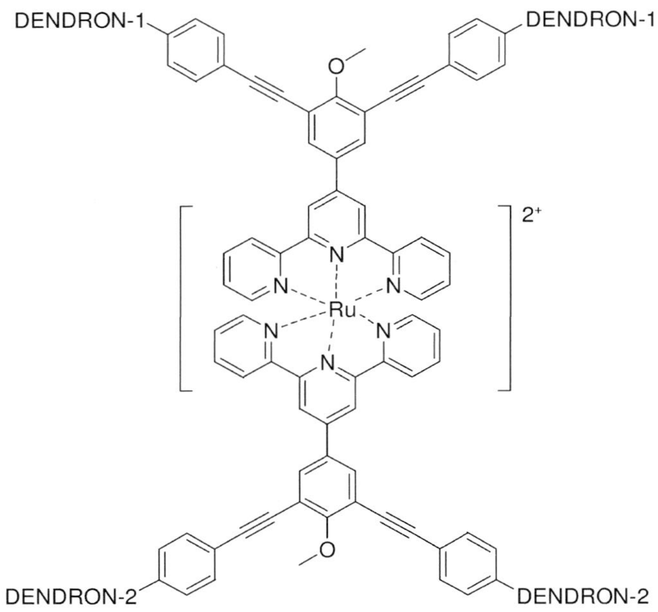

Figure 6.4 Diederich et al.'s amphiphilic "vectors" for gene delivery.

Research on related star-like macromolecules with terpyridine metal complexes in the outer sphere has been conducted by diverse groups [17, 31, 44–54]. In 1996, Constable et al. [48, 55] reported the formation of a star-like macromolecule with 18 Ru(II) centers by reaction of hexakis(bromomethyl)benzene and six dendrons, each containing three Ru complexes (Figure 6.5); the related *tris*-armed relative was also reported [56]. A pentaerythritol-based metallodendrimer [50] was also tested as a mediator for the electrochemical oxidation of methionine, cystine, and As(III) [57]. In the evaluation of surfaces, Constable's pentaerythritol-cored metallodendrimer and a Dawson-type phosphotungstate, $[P_2W_{18}O_{62}]^{6-}$ (P_2W_{18}) were coated on an electrode surface and shown to be "highly organized, bifunctional catalytic systems with the stability needed for practical application[s]" [58].

In 1996, Marvaud and Astruc [17] reported aromatic metallostars containing hexa(terpyridine) branches with and without a central $Fe(\eta^5\text{-}C_5H_5)^+$ core and a $[Ru(tpy)_2]^{2+}$ surface.

Constable et al. [47] also published a "first generation" star-like system based on pentaerythritol-bearing pendant [Ru(4'-(2-(tert-butyldimethylsilyl)-1,2-carbaboranyl)-tpy)(tpy)]$^{2+}$ complexes. MALDI-TOF-MS was demonstrated to be a very useful tool for the characterization of this high-molecular-weight, octa-charged system, revealing a signal at m/z = 3420 Dalton, which was assigned to the loss of all 8 counterions. More recently, the pentaerythritol core was used for assembling a novel 2nd generation hexadecanuclear metallo-supramolecular system [50].

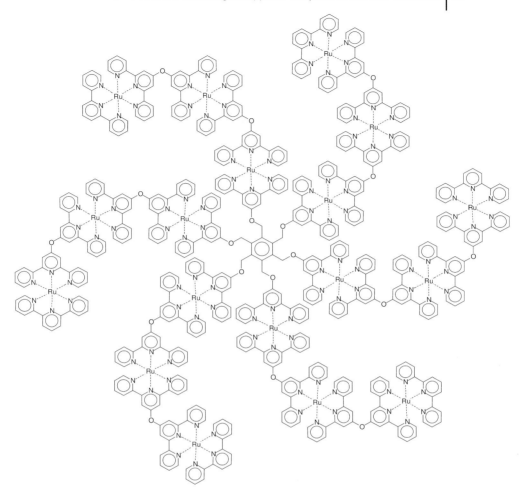

Figure 6.5 A metallomer possessing 18 $[Ru(tpy)_2]^{2+}$ centers [47, 48].

6.3
Dendrimers Containing *Bi*sterpyridine Complexes as Non-Core Connectors

Tailor-made isomeric metallodendrimers were synthesized by Newkome et al. [52]. Both dendrimers shown in Figure 6.6 have identical molecular formulae $(C_{597}H_{880}F_{48}N_{52}O_{136}P_8Ru_4)$ and a molecular mass of 12 526 Dalton. By controlling the design of the macromolecular architecture, the nanoscopic tetrahedral geometry of methane was mimicked. While IR, UV, NMR, and MALDI-TOF-MS experiments revealed very similar results, electrochemical studies indicated that internal densities and void regions differed greatly. The voltammograms of the two compounds are similar and exhibit two quasi-reversible waves at negative

connectivity and internal counterions but also bearing surface adamantane groups capable of host-guest chemistry. The different metallodendrimers were subsequently reacted with cyclodextrin, which formed complexes with the surface adamantane termini leading to a supramolecular assembly (Scheme 6.3).

Introducing chirality as well as photoactive $[Ru(tpy)_2]^{2+}$ complexes into the dendritic sphere is of interest for the creation of macromolecules with new optical properties. Lin et al. [64] functionalized Fréchet-type benzyl bromide dendrons with a binaphthyl-terpyridine, which were subsequently complexed to tetrakis(2,2':6',2''-terpyridin-4'-oxymethyl)methanes by applying Ru(III)/Ru(II) chemistry (Figure 6.9). CD spectroscopy revealed an enantiomerically pure compound. It was found that the $[Ru(tpy)_2]^{2+}$ complexes quenched the fluorescence of the binaphthyl as well as dimethoxybenzyl fluorophores, which was explained by the

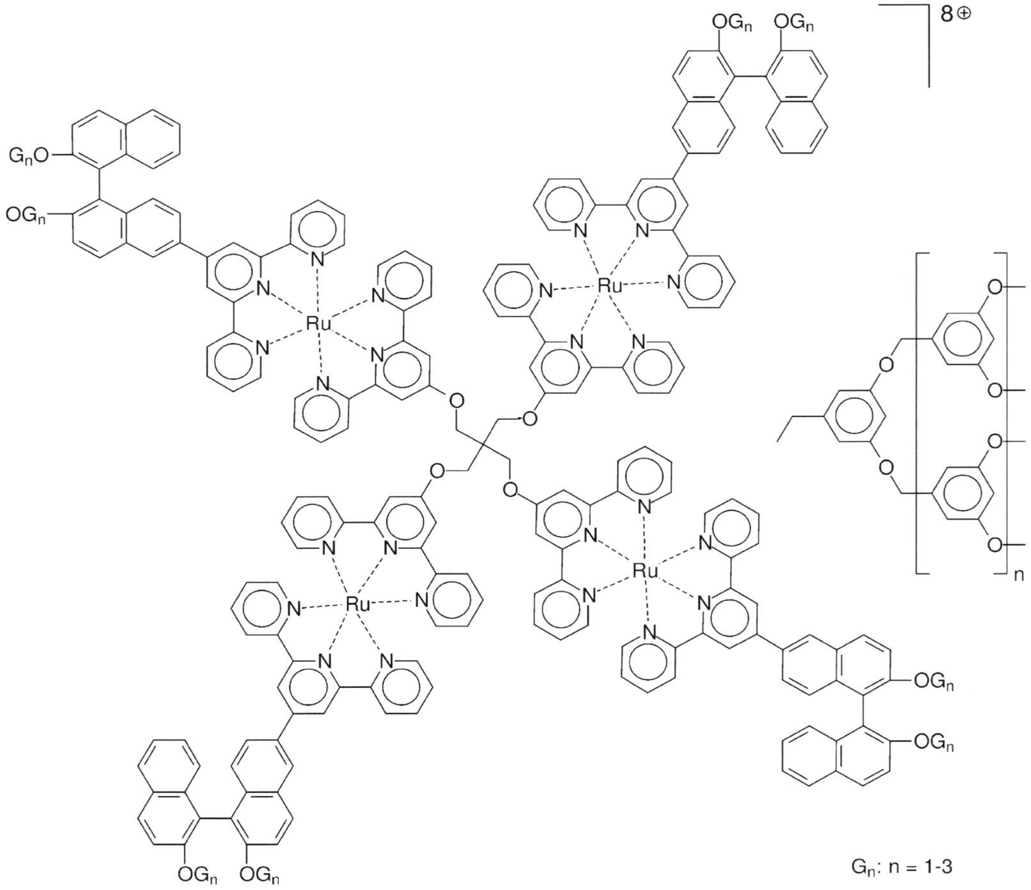

Figure 6.9 Fréchet-type dendrons containing terpyridines which are connected to the core with $[Ru(tpy)_2]^{2+}$ centers [64].

quenching of the ^{3}MLCT states by low-lying metal-centered states. It is believed that the intersystem crossing from the π^* bands of the peripheral fluorophore to the ^{3}MLCT states is highly efficient, resulting in an emission of the π^* system to be quenched. Only for the highest generation dendrimer could a weak luminescence be observed, supporting distance dependence of the quenching efficiency. In a similar approach, these authors synthesized a metallodendrimer with the same core structure but lacking the dimethoxybenzyl dendron arms [65]. The circular dichroism (CD) spectra for both of these dendrimers showed three Cotton effects and were very similar to the spectra of the *bis*-naphthyl ligands, thus showing no sign of newly formed chirality.

Osawa et al. [66] have created a series of rigid dendritic nano-sized [Ru(tpy)$_2$]$^{2+}$ complexes based on a simple Pd-mediated, one-step coupling of polypyridine Ru complexes; this procedure generated a family of cationic complexes with 6, 12, and 18 peripheral [Ru(tpy)$_2$]$^{2+}$ units with the largest possessing a diameter of 9 nm with 38 positive charges (Figure 6.10).

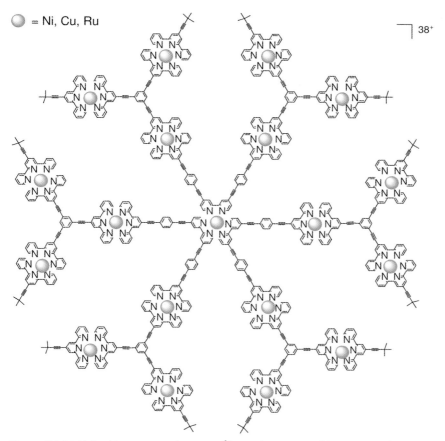

Figure 6.10 Rigid dendritic nano-sized [Ru(tpy)$_2$]$^{2+}$ complexes prepared by Osawa et al. [66].

More complex heteroleptic supramolecular metallodendrimers were synthesized by Newkome et al. [67]. They applied mixtures of differently functionalized [Ru(R-tpy)(Cl)$_3$] dendrons with an octaterpyridinyl polyamide dendritic core in order to achieve a combinatorial-type approach based on terpyridine-Ru(III)/Ru(II) chemistry (Scheme 6.4). Initially, the terpyridine-coated PPI dendrimer was separately treated with each of the three [Ru(4'-R-tpy)(Cl)$_3$] compounds, which led to dendrimers with different ethereal functions at the periphery. Subsequently, mixtures of the three different [Ru(4'-R-tpy)(Cl)$_3$] ligands were reacted with the terpyridine-coated dendrimer. Upon comparison of the ^{13}C NMR spectrum of the different species, it could be shown that the three different functionalities were successfully complexed to the dendritic core. Furthermore, the selective hydrolysis of the peripheral *tert*-butyl esters as well as the selective debenzylation of the benzyl ether functionalities of the other complexed ligand were demonstrated. This combinatorial strategy combined with orthogonal decoupling

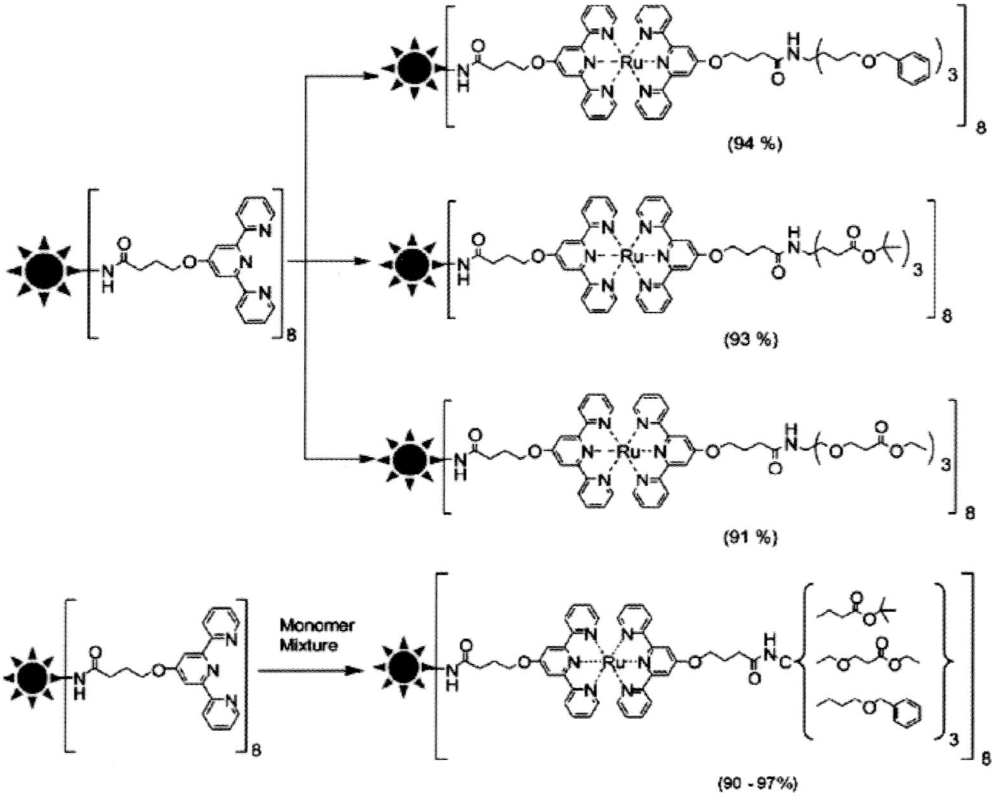

Scheme 6.4 Synthesis of polyamide terpyridine core homogeneous (top) and heterogeneous (bottom) metallodendrimers [67]. (Reproduced with permission from [67], © 2003 Elsevier).

techniques led to systems with latent or masked regions of reactivity that can be accessed and addressed specifically at a desired time in generational construction. Therefore, an advantageous flexibility between that of a completely directed approach, whereby precise control over monomer attachment was maintained, and that of random, uncontrolled mixed monomer attachment to reactants on the surface was achieved. Differential scanning calorimetry and thermogravimetric analysis on the family of metallodendrimers were used to ascertain their thermal behavior, glass-transition temperatures, and decomposition kinetics and temperatures; no synergy effects were determined for the heterogeneous series, which was in contrast to the corresponding homogeneously coated materials [68].

6.4
Dendrimers Containing *Bisterpyridine* Complexes at the Surface

Constable et al. [49, 70–73] has shown that it is possible to build star-like molecules possessing an $[Fe(bpy)_3]^{2+}$ or $[Co(bpy)_3]^{2+}$ core and $[Ru(tpy)_2]^{2+}$ complexes in the outer sphere – a convergent-type approach by connecting three bipyridine-functionalized arms through the formation of a *tris*bipyridine core. NMR spectroscopy and MALDI-TOF-MS were used to characterize the metallostar. In the case with an Fe(II) core, the most intensive peak in the MALDI-TOF-MS was exhibited at $m/z = 6330$ Dalton, indicating the loss of 3 PF_6^- counterions. For the metallostar with Co(II) core, the loss of 2–4 counterions was detected.

Kimura et al. [44] reported the use of up to the 3rd generation, and Abruña et al. reported [45, 46, 53, 74] the use of up to the 4th generation poly(amidoamine) [PAMAM] dendrimers, possessing an ethylenediamine core with surface $[Ru(tpy)_2]^{2+}$ moieties. The 4th generation dendrimer (Figure 6.11) contains 64 $[Ru(tpy)_2]^{2+}$ complexes and thus 128 positive charges and 128 counterions. Both groups used peptide coupling procedures for synthesis. Abruña et al. demonstrated that the interfacial reaction of (a) the terpyridine-pendant dendrimers and a bridging ligand 1,4-*bis*[4,4″-*bis*(1,1-dimethylethyl)terpyridin-4′-yl]benzene dissolved in CH_2Cl_2 and (b) an aqueous solution of metal ions gave rise to ordered films on highly oriented pyrolytic graphite. STM investigations of these films demonstrated highly ordered hexagonal 2-D domains (in this case dend-8-tpy/ Fe^{2+}) [46].

Hong and Murfee [75–78] recently generated dendrimers up to the 3rd generation based on an octa(diphenylphosphino)-functionalized silsesquioxane core. The 3rd generation species was constructed from 8 dendrons, each possessing 4 terpyridine molecules, to which $[Ru(tpy)(bpy)_2]^{2+}$ complexes capped the surface ligands (Figure 6.12). The molecule thus had 64 positive charges at its outer sphere and 64 hexafluorophosphate counterions. The dendrimer was characterized by photophysical and electrochemical methods. Quantum yields were determined to range from 2.1×10^{-2} to 1.1×10^{-2} depending on the dendrimer's generation. Ruthenium(II)-based excited-state lifetimes were 605, 890, and 880 ns, again depending on the generation; the emission wavelength was 610 nm.

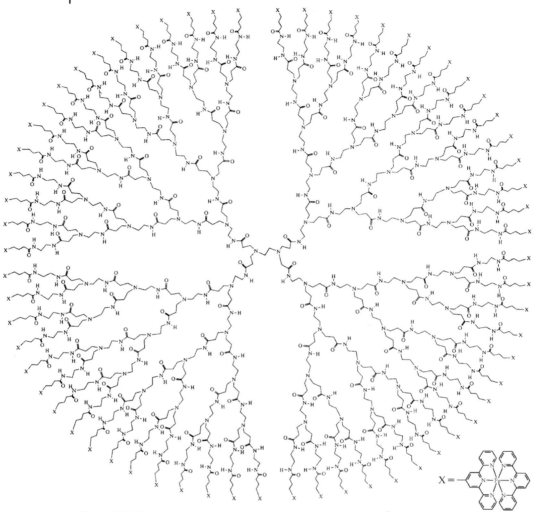

Figure 6.11 The 4th generation of a dendrimer with pendant [M(tpy)$_2$]$^{2+}$ complexes (X) [45, 46]. (Reprinted with permission from [53], © 1999 American Chemical Society).

Kim et al. [79] reported the synthesis and characterization of three generations of carbosilanes possessing a surface of 4, 8, and 16 terpyridines, respectively (Figure 6.13). These dendrimers were also suitable for complexation to other terpyridines using the traditional Ru(III)/Ru(II) strategy. Characterization was carried out by MALDI-TOF-MS, NMR, and UV/Vis absorption. Generally, dendritic carbosilanes are of special interest because they are chemically inert and fluid at high molecular weight at the higher generations. Therefore, the combination with [Ru(tpy)$_2$]$^{2+}$ complexes could lead to unique materials with special photophysical properties.

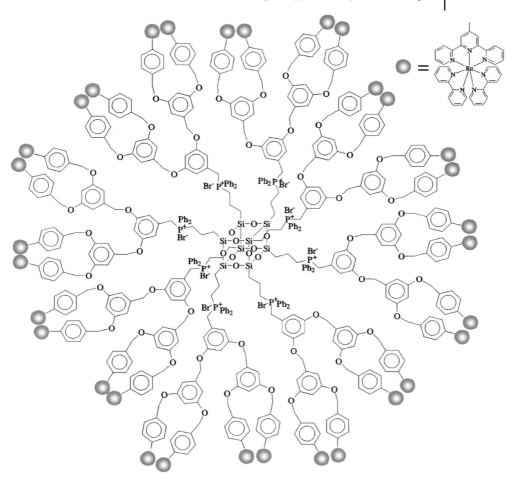

Figure 6.12 Third generation dendrimer based on a silsesquioxane core [75–78].

Large polymeric systems or aggregates, such as micelles and cross-linked resin-type structures, offer interesting alternatives to homogeneous, solution chemistry, e.g., asymmetric catalytic reductions of carbonyl bonds [80] or different solid-phase synthesis approaches demonstrated first in 1965 by Merrifield, who anchored reagents to insoluble supports for purposes of solid-phase peptide synthesis (SPPS) [81].

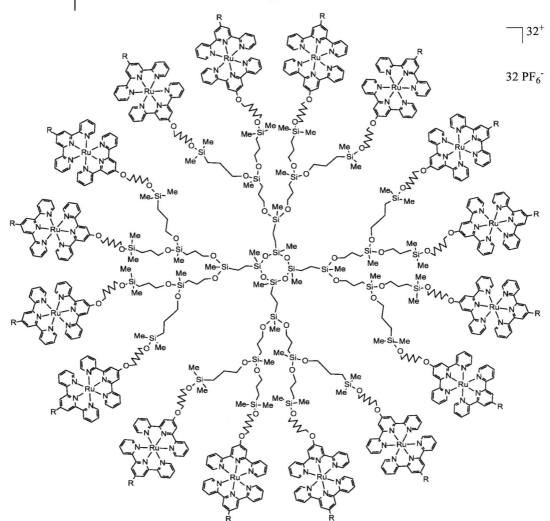

Figure 6.13 3rd-Generation carbosilane metallodendrimer with surface
$[Ru(tpy)_2]^{2+}$ moieties. (Reproducedwith permission from [79], © 2003 Elsevier).

6.5
Micelles Composed of Terpyridine-Complex-Containing Polymers

Only Schubert et al. have so far reported any investigations into micellar aggregates incorporating terpyridine complexes. Micelles were formed using the systems reported above, which consist of a hydrophilic/hydrophobic block copolymer linked by $[Ru(tpy)_2]^{2+}$ complexation. As the first example, $PS_{20}-[Ru(tpy)_2]^{2+}-PEO_{70}$ represents an amphiphilic block copolymer which was found to aggregate to micelles in water (Figure 6.14) [82]. Concerning the stability of these micelles in terms of $[Ru(tpy)_2]^{2+}$ connectivity, they were found to be stable upon variation of temperature (20–70 °C), ionic strength (pure water – 1 M NaCl) and pH (0–14); however, there were significant influences concerning D_h.

In a different system including the soft poly(ethylene-*co*-butylene) core, as opposed to the glassy PS core [83], the Ru(II) complexes of the $PEB_{70}-[Ru(tpy)_2]^{2+}-PEO_{70}$ could be opened by the addition of a competitive ligand, such as HEEDTA. The solution turned from red to colorless, and DLS measurements confirmed the existence of objects with $D_h = 13$ nm, which were thought to be PEB cores dispersed in water. Moreover, more complex systems using $PS_{32}-P2VP_{13}-[Ru(tpy)_2]^{2+}-PEG_{70}$ block copolymers have also been prepared [84].

As well as linear block copolymers, amphiphilic graft copolymers based on a terpyridine-functionalized poly(methyl methacrylate) and poly(ethylene glycol) side chains were converted into spherical micelles. These showed higher polydispersity than that of the block copolymer micelles [85].

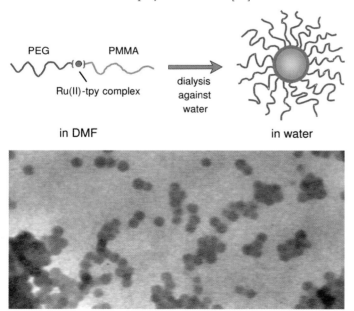

Figure 6.14 Top: schematic overview of the micelle formation.
Bottom: TEM image of the micelles.

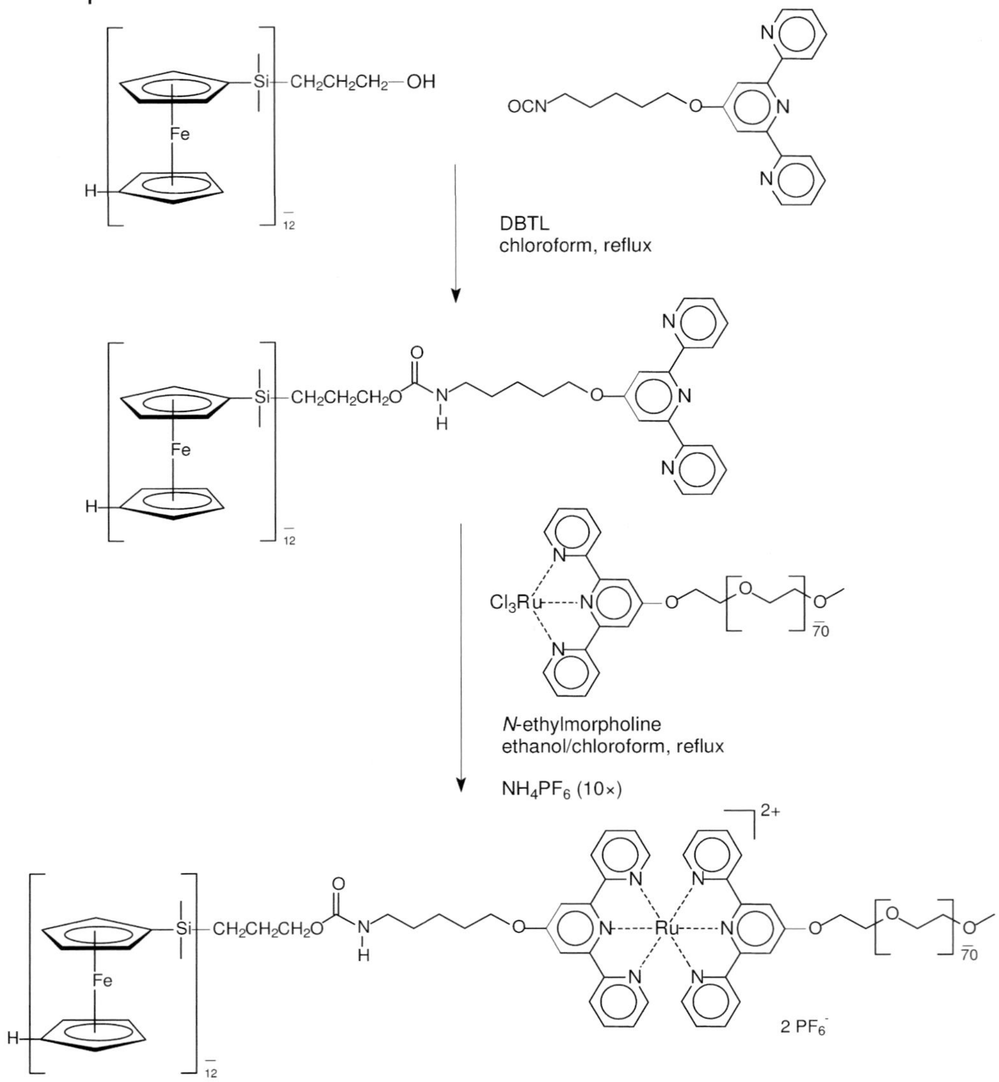

Scheme 6.5 Schematic representation of the synthesis of the poly(ferrocenylsilane)-poly(ethylene glycol) block copolymer [86].

As well as spherical micelles, cylindrical micellar aggregates could also be obtained from supramolecular block copolymers. Recently, the preparation of an AB block copolymer consisting of a poly(ethylene glycol) and a poly(ferrocenyl-silane) with a block ratio of 6 : 1 was reported (Scheme 6.5). Applying similar techniques, micelles were obtained; TEM (Figure 6.15) and AFM imaging revealed rod-like cylindrical micellar aggregates [86].

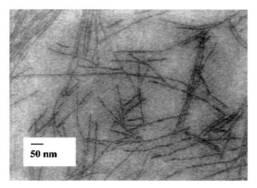

Figure 6.15 TEM image of the cylindrical micelles.
(Reprinted with permission from [86]).

6.6
Resins and Beads Modified with Terpyridine

Up to now, limited reports on terpyridine-functionalized polymeric resins or beads
have appeared in the literature, at least in comparison with those possessing
functionalized bipyridines [87]. The main applications of such species lie in the
area of heterogeneous catalysis.

Yoo et al. [88] reported the reaction of functionalized poly(chloromethylstyrene-
co-divinylbenzene) (PCD) with a 4′-(4-hydroxyphenyl)terpyridine (Scheme 6.6).
The product was complexed with $Fe(ClO_4)_3$ to form the Fe(III)-terpyridine *mono*-
complexes. The formation of *bis*-complexes seemed unlikely in that the terpyridine
ligand was attached in a rather sterically confined, rigid manner. This material
was then used to catalyze the ring-opening of different epoxides with methanol
and water; almost quantitative conversions after short reaction times at room
temperature were demonstrated.

Buchmeiser et al. [89] also investigated the possibility of developing a polymer-
supported catalyst; in their case, heterogeneous Atom Transfer Radical Polymerization (ATRP) was utilized. However, the terpyridine–Cu(I) grafted PS-DVB
(2% cross-linking) did not lead to the isolation of any PS.

The loading of different metal-ions [Fe(II), Co(II), Cu(II), Ru(III) or Ni(II)] on
terpyridine-functionalized TentaGel (PS/PEO) microbeads ($d = 20\ \mu m$) was
investigated by Schubert et al. [90, 91] (Scheme 6.7 and Figure 6.16). UV/Vis
spectroscopic measurements of the suspensions of these beads led to the
characteristic absorption bands known from the free terpyridine metal complexes;
complexation also became apparent from the resultant coloration of the material.
Furthermore, the loadings of these beads could be investigated by AAS; loading
rates were found to be in accordance with quantitative *mono*- and *bis*-complexation
of the terpyridine on the microbeads. Application of Ru(III)/Ru(II) chemistry led
to the functionalization of the terpyridine moieties with an anthracene-functiona-
lized terpyridine.

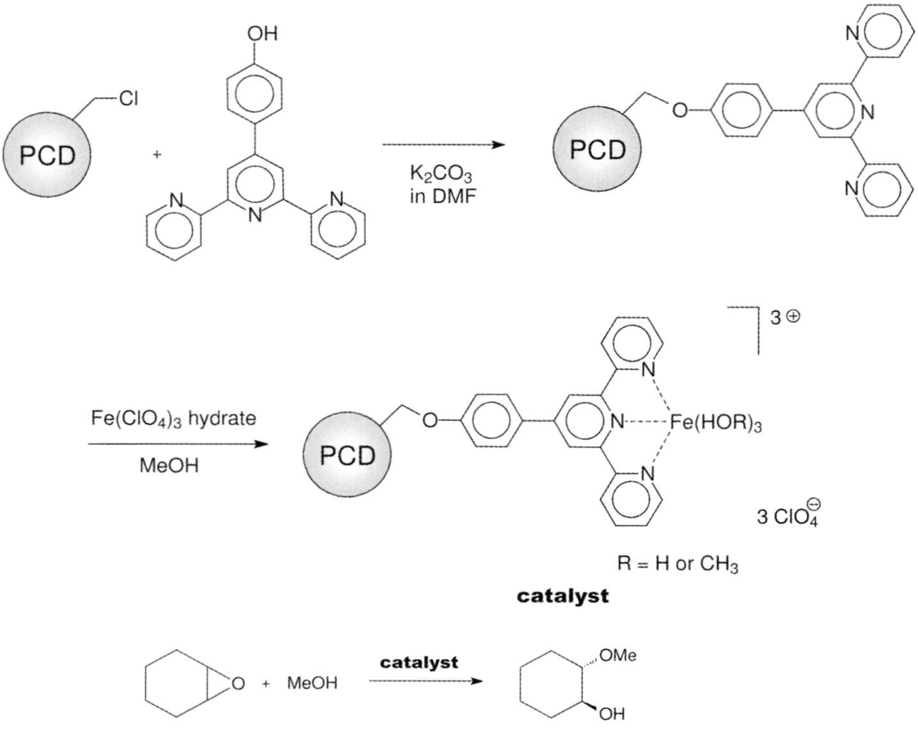

Scheme 6.6 Top: synthesis of polymer-supported poly(chloromethyl-styrene-co-divinylbenzene)(PCD)-terpyridine-Fe(III) catalyst.
Bottom: hydrolysis of epoxides using PCD-terpyridine-Fe(III) catalyst in a mixture of acetone/H_2O (8 : 2 v/v) at room temperature [88].

substrate	time (h)	conversion yield (%)
cyclohexene oxide	2	> 99
cyclopentene oxide	32	> 99
styrene oxide	48	> 99
1-hexene oxide	120	> 99

For the characterization of these terpyridines attached to solid polystyrene supports, Heinze et al. [92] developed a new mass-spectrometric technique in order to detect the attached terpyridine moieties. Cross-linked polystyrene was first modified with a silyl-ether linker, and this was then reacted with 4'-(4-hydroxyphenyl)terpyridine. Dry samples of the resulting material were ground to a fine powder before introduction into the EI mass spectrometer. This method led to the detection of fragments which could be distinguished from unbound material.

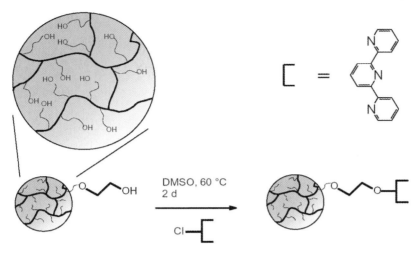

Scheme 6.7 Functionalization of TentaGel® microbeads with terpyridine moieties [90, 91].

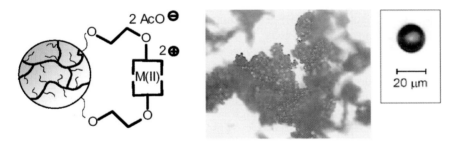

Figure 6.16 Complexation of terpyridine-functionalized TentaGel® microbeads with metal ions, including photograph of $[Fe(tpy)_2]^{2+}$-complexed beads (right) [90]. (Reprinted with permission from [91]).

References

1 G. R. Newkome, C. N. Moorefield, F. Vögtle, *Dendrimers and Dendrons: Concepts, Syntheses, Applications,* Wiley-VCH, Weinheim, Germany, **2001**.

2 R. Arshady, *Microspheres, Microcapsules and Liposomes* **2002**, *5*, 1–29.

3 M. Venturi, S. Serroni, A. Juris, S. Campagna, V. Balzani, *Top. Curr. Chem.* **1998**, *197*, 193–228.

4 D. Astruc, J.-C. Blais, E. Cloutet, L. Djakovitch, S. Rigaut, J. Ruiz, V. Sartor, C. Valério, *Top. Curr. Chem.* **2000**, *210*, 229–259.

5 D. Astruc, F. Chardac, *Chem. Rev.* **2001**, *101*, 2991–3023.

6 D. Astruc, J.-C. Blais, M.-C. Daniel, V. Martinez, S. Nlate, J. Ruiz, *Macromol. Symp.* **2003**, *196*, 1–25.

7 R. M. Crooks, M. Zhao, L. Sun, V. Chechik, L. K. Yeung, *Acc. Chem. Res.* **2001**, *34*, 181–190.

8 A. W. Kleij, A. Ford, J. T. B. H. Jastrzebski, G. van Koten, *Dendritic Polymer Applications: Catalysts*, in *Dendrimers and Other Dendritic Polymers*, John Wiley & Sons, Ltd., West Sussex, UK, **2001**, pp. 485–514.

9 G. E. Oosterom, J. N. H. Reek, P. C. J. Kamer, *Angew. Chem. Int. Ed.* **2001**, *40*, 1828–1849.

10 V. Balzani, P. Ceroni, A. Juris, M. Venturi, S. Campagna, F. Puntoriero, S. Serroni, *Coord. Chem. Rev.* **2001**, *219–221*, 545–572.

11 S. Campagna, C. Di Pietro, F. Loiseau, B. Maubert, N. McClenaghan, R. Passalacqua, F. Puntoriero, V. Ricevuto, S. Serroni, *Coord. Chem. Rev.* **2002**, *229*, 67–74.

12 P. Ceroni, V. Vicinelli, M. Maestri, V. Balzani, S.-K. Lee, J. van Heyst, M. Gorka, F. Vögtle, *J. Organomet. Chem.* **2004**, *689*, 4375–4383.

13 M.-S. Choi, T. Yamazaki, I. Yamazaki, T. Aida, *Angew. Chem. Int. Ed.* **2004**, *43*, 150–158.

14 A. Dirksen, L. De Cola, *C. R. Chimie* **2003**, *6*, 873–882.

15 H. Frauenrath, *Prog. Polym. Sci.* **2005**, *30*, 325–384.

16 G. R. Newkome, E. He, C. N. Moorefield, *Chem. Rev.* **1999**, *99*, 1689–1746.

17 V. Marvaud, D. Astruc, *Chem. Commun.* **1997**, *21*, 773–774.

18 C. Gorman, *Adv. Mater.* **1998**, *10*, 295–309.

19 F. J. Stoddart, T. Welton, *Polyhedron* **1999**, *18*, 3575–3591.

20 B. Alonso, E. Alonso, D. Astruc, J.-C. Blais, L. Djakovitch, J.-L. Fillaut, S. Nlate, F. Moulines, S. Rigaut, J. Ruiz, C. Valério, *Dendrimers containing ferrocenyl or other transition-metal sandwich groups*, in *Advances in Dendritic Macromolecules*, Elsevier Science Ltd., Kidlington, Oxford UK, **2002**, pp. 89–127.

21 P. A. Chase, R. J. M. K. Gebbink, G. van Koten, *J. Organomet. Chem.* **2004**, *689*, 4016–4054.

22 I. Cuadrado, M. Morán, C. M. Casado, B. Alonso, J. Losada, *Coord. Chem. Rev.* **1999**, *193–195*, 395–445.

23 K. Onitsuka, S. Takahashi, *Top. Curr. Chem.* **2003**, *228*, 39–63.

24 J. N. H. Reek, D. de Groot, G. E. Oosterom, P. C. J. Kamer, P. W. N. M. van Leeuwen, *Rev. Mol. Biotech.* **2002**, *90*, 159–180.

25 M. Venturi, P. Ceroni, *C. R. Chimie* **2003**, *6*, 935–945.

26 C. N. Moorefield, G. R. Newkome, *C. R. Chimie* **2003**, *6*, 715–724.

27 L. J. Twyman, A. S. H. King, I. K. Martin, *Chem. Soc. Rev.* **2002**, *31*, 69–82.

28 F. Aulenta, W. Hayes, S. Rannard, *Eur. Polym. J.* **2003**, *39*, 1741–1771.

29 U. Boas, P. M. H. Heegaard, *Chem. Soc. Rev.* **2004**, *33*, 43–63.

30 A. K. Patri, I. J. Majoros, J. R. Baker Jr., *Curr. Opin. Chem. Biol.* **2002**, *6*, 466–471.

31 G. R. Newkome, F. Cardullo, E. C. Constable, C. N. Moorefield, A. M. W. C. Thompson, *Chem. Commun.* **1993**, 925–927.

32 E. C. Constable, M. D. Ward, *J. Chem. Soc., Dalton Trans.* **1990**, 1405–1409.

33 G. R. Newkome, C. N. Moorefield, R. Günther, G. R. Baker, *Polym. Prepr.* **1995**, *36*, 609–610.

34 G. R. Newkome, R. Güther, C. N. Moorefield, F. Cardullo, L. Echegoyen, E. Pérez-Cordero, H. Luftmann, *Angew. Chem.* **1995**, *107*, 2159–2162.

35 G. R. Newkome, R. Güther, C. N. Moorefield, F. Cardullo, L. Echegoyen, E. Pérez-Cordero, H. Luftmann, *Angew. Chem. Int. Ed.* **1995**, *34*, 2023–2026.

36 H.-F. Chow, I. Y.-K. Chan, C. C. Mak, M.-K. Ng, *Tetrahedron* **1996**, *52*, 4277–4290.

37 H.-F. Chow, I. V.-K. Chan, C. C. Mak, *Tetrahedron Lett.* **1995**, *36*, 8633–8636.

38 H. F. Chow, I. Y. K. Chan, D. T. W. Chan, R. W. M. Kwok, *Chem. Eur. J.* **1996**, *2*, 1085–1091.

39 R. L. C. Lau, T. W. D. Chan, I. Y. K. Chan, H. F. Chow, *Eur. Mass Spec.* **1995**, *1*, 371–380.

40 H. F. Chow, I. Y. K. Chan, P. S. Fung, T. K. K. Mong, M. F. Nongrum, *Tetrahedron* **2001**, *57*, 1565–1572.

41 E. C. Constable, C. E. Housecroft, M. Neuburger, S. Schaffner, L. J. Scherer, *Dalton Trans.* **2004**, 2635–2642.

42 M. Kimura, T. Shiba, T. Muto, K. Hanabusa, H. Shirai, *Chem. Commun.* **2000**, 11–12.

43 D. Joester, V. Gramlich, F. Diederich, *Helv. Chim. Acta* **2004**, *87*, 2896–2918.

44 M. Kimura, K. Mizuno, T. Muto, K. Hanabusa, H. Shirai, *Macromol. Rapid Commun.* **1999**, *20*, 98–102.

45 G. D. Storrier, K. Takada, H. D. Abruña, *Langmuir* **1999**, *15*, 872–884.

46 D. J. Diaz, G. D. Storrier, S. Bernhard, K. Takada, H. D. Abruña, *Langmuir* **1999**, *15*, 7351–7354.

47 D. Armspach, M. Cattalini, E. C. Constable, C. E. Housecroft, D. Phillips, *Chem. Commun.* **1996**, 1823–1824.

48 E. C. Constable, P. Harverson, *Inorg. Chim. Acta* **1996**, *252*, 9–11.

49 E. C. Constable, *Chem. Commun.* **1997**, 1073–1080.

50 E. C. Constable, C. E. Housecroft, M. Cattalini, D. Phillips, *New J. Chem.* **1998**, 193–200.

51 G. R. Newkome, E. He, *J. Mater. Chem.* **1997**, *7*, 1237–1244.

52 G. R. Newkome, E. He, L. A. Godinez, *Macromolecules* **1998**, *31*, 4382–4386.

53 K. Takada, G. D. Storrier, M. Morán, H. D. Abruña, *Langmuir* **1999**, *15*, 7333–7339.

54 G. R. Newkome, E. He, L. A. Godínez, G. R. Baker, *Chem. Commun.* **1999**, 27–28.

55 E. C. Constable, P. Harverson, *Chem. Commun.* **1996**, 33–34.

56 E. C. Constable, P. Harverson, J. J. Ramsden, *Chem. Commun.* **1997**, 1683–1684.

57 S. D. Holmstrom, J. A. Cox, *Anal. Chem.* **2000**, *72*, 3191–3195.

58 L. Cheng, J. A. Cox, *Chem. Mater.* **2002**, *14*, 6–8.

59 P. J. Dandliker, F. Diederich, J.-P. Gisselbrecht, A. Louati, M. Gross, *Angew. Chem. Int. Ed.* **1995**, *34*, 2725–2728.

60 C. B. Gorman, B. L. Parkhurst, W. Y. Su, K.-Y. Chen, *J. Am. Chem. Soc.* **1997**, *119*, 1141–1142.

61 G. R. Newkome, V. V. Narayanan, L. Echegoyen, E. Pèrez-Cordero, H. Luftmann, *Macromolecules* **1997**, *30*, 5187–5191.

62 G. R. Newkome, E. He, L. A. Godínez, G. R. Baker, *J. Am. Chem. Soc.* **2000**, *122*, 9993–10006.

63 G. R. Newkome, H. J. Kim, K. H. Choi, C. N. Moorefield, *Macromolecules* **2004**, *37*, 6268–6274.

64 H. Jiang, S. J. Lee, W. Lin, *J. Chem. Soc., Dalton Trans.* **2002**, 3429–3433.

65 H. Jiang, J. Lee Suk, W. Lin, *Org. Lett.* **2002**, *4*, 2149–2152.

66 M. Osawa, M. Hoshino, S. Horiuchi, Y. Wakatsuki, *Organometallics* **1999**, *18*, 112–114.

67 G. R. Newkome, K. S. Yoo, S.-H. Hwang, C. N. Moorefield, *Tetrahedron* **2003**, *59*, 3955–3964.

68 S.-H. Hwang, K. S. Yoo, C. N. Moorefield, G. R. Newkome, *J. Polym. Sci., Part B: Polym. Phys.* **2003**, *42*, 1487–1495.

69 G. R. Newkome, K. S. Yoo, H. J. Kim, C. N. Moorefield, *Chem. Eur. J.* **2003**, *9*, 3367–3374.

70 E. C. Constable, P. Harverson, M. Oberholzer, *J. Chem. Soc., Chem. Commun.* **1996**, 1821–1822.

71 E. C. Constable, C. E. Housecroft, *Chimia* **1999**, *53*, 187–191.

72 E. C. Constable, P. Harverson, *Polyhedron* **1999**, *18*, 1891–1901.

73 E. C. Constable, C. E. Housecroft, I. Poleschak, *Inorg. Chem. Commun.* **1999**, *2*, 565–568.

74 J. I. Goldsmith, K. Takada, H. D. Abruña, *J. Phys. Chem. B* **2002**, *106*, 8504–8513.

75 H. J. Murfee, B. Hong, *Polym. Prepr.* **2000**, *41*, 431–432.

76 B. Hong, T. P. S. Thoms, H. J. Murfee, M. J. Lebrun, *Inorg. Chem.* **1997**, *36*, 6146–6147.

77 H. J. Murfee, B. Hong, *Polym. Prepr.* **1999**, *40*, 412–413.

78 H. J. Murfee, T. P. S. Thoms, J. Greaves, B. Hong, *Inorg. Chem.* **2000**, *39*, 5209–5217.

79 C. Kim, H. Kim, *J. Organomet. Chem.* **2003**, *673*, 77–83.

80 C. Saluzzo, R. ter Halle, F. Touchard, F. Fache, E. Schulz, M. Lemaire, *J. Organomet. Chem.* **2000**, *603*, 30–39.

81 R. B. Merrifield, *Science* **1965**, *150*, 178–185.

82 J.-F. Gohy, B. G. G. Lohmeijer, U. S. Schubert, *Macromolecules* **2002**, *35*, 4560–4563.

83 J.-F. Gohy, B. G. G. Lohmeijer, U. S. Schubert, *Macromol. Rapid Commun.* **2002**, *23*, 555–560.

84 J.-F. Gohy, B. G. G. Lohmeijer, S. K. Varshney, B. Decamps, E. Leroy, S. Boileau, U. S. Schubert, *Macromolecules* **2002**, *35*, 9748–9755.

85 J.-F. Gohy, H. Hofmeier, A. S. Alexeev, U. S. Schubert, *Macromol. Chem. Phys.* **2003**, *204*, 1524–1530.

86 J.-F. Gohy, B. G. G. Lohmeijer, A. S. Alexeev, X.-S. Wang, I. Manners, M. A. Winnik, U. S. Schubert, *Chem. Eur. J.* **2004**, *10*, 4315–4323.

87 U. S. Schubert, C. Eschbaumer, *Angew. Chem. Int. Ed.* **2002**, *41*, 2892–2926.

88 D.-W. Yoo, S.-K. Yoo, C. Kim, J.-K. Lee, *J. Chem. Soc., Dalton Trans.* **2002**, 3931–3932.

89 R. Kröll, C. Eschbaumer, U. S. Schubert, M. R. Buchmeiser, K. Wurst, *Macromol. Chem. Phys.* **2001**, *202*, 645–653.

90 P. R. Andres, U. S. Schubert, *Polym. Mater. Sci. Eng.* **2003**, *88*, 356–357.

91 U. S. Schubert, A. Alexeev, P. R. Andres, *Macromol. Mater. Eng.* **2003**, *288*, 852–860.

92 K. Heinze, U. Winterhalter, T. Jannack, *Chem. Eur. J.* **2000**, 6, 4203–4210.

7
Surfaces Modified with Terpyridine Metal Complexes

7.1
Introduction

The fundamental understanding of molecular scale, ordered nanostructures on various surfaces is a rapidly growing field, partly because of recent technical advances [1, 2]. Also, terpyridine complexes play an ever-increasing role in the area of functional applications, such as solar cell devices or electrode catalysis. Furthermore, such easily detectable and multi-functional entities are critically important in obtaining the desired level of deep-seated understanding of the self-organization of organic or inorganic-organic hybrid materials on a variety of surfaces.

For example, the creation of new functional materials was recently described in which a bipyridine ruthenium complex was homogeneously dispersed in a poly(phenylene vinylene) (PPV) derivative in order to act as an electroluminescent entity for reversible switching between red and green emission [3]. It was demonstrated that if a forward bias voltage was applied, the excited state of the ruthenium complex was populated and the characteristic red emission was observed, whereas, on reversing the bias, the lowest excited singlet state of the polymer host was populated, with consequent emission of green light. This simple example demonstrates the potential uses of such polypyridine-metal complexes in combination with polymeric materials and surfaces.

7.2
Assemblies and Layers

Research into the molecular-level modification of surface properties has been increasing ever since Binnig and Rohrer in the mid-1980s invented the pivotal scanning-probe techniques, such as STM (scanning tunneling microscopy) or AFM (atomic force microscopy); however, there is still much to learn in terms of predesigning order and orientation of substances on a given surface. Metallo-supramolecular structures especially add a whole range of possibilities, not least because of possible interactions between the complexed metal and the metal

Modern Terpyridine Chemistry. U. S. Schubert, H. Hofmeier, G. R. Newkome
Copyright © 2006 WILEY-VCH Verlag GmbH & Co. KGaA, Weinheim
ISBN: 3-527-31475-X

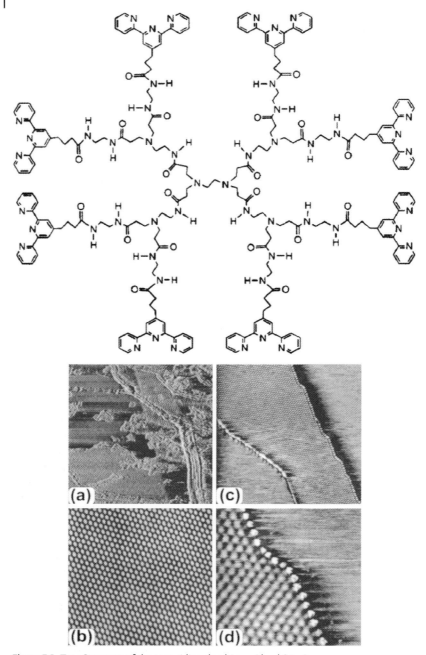

Figure 7.1 Top: Structure of the terpyridine dendrimer (dend-8-tpy).
Bottom: Unfiltered images of dend-8-[Fe(tpy)$_2$]$^{2+}$ on HOPG:
(a) 550 × 550 nm, (b) 200 × 200 nm, (c) 304 × 304 nm, (d) 69 × 69 nm.
(Reprinted with permission from [5], © 1999 American Chemical Society).

surface. Following the theme of this book, there has been growing interest in the investigation of adsorbed ordered structures on surfaces containing $[M(tpy)_2]^{2+}$ complexes. For an overview of layer-by-layer self-assemblies containing terpyridine complexes, the reader is referred to a review (see [4]).

Abruña et al. recently described the synthesis of chiral and dendritic multi-terpyridine molecules, which upon complexation with Fe(II) or Co(II) and subsequent deposition on a surface, led to well-ordered 2-dimensional arrays [5]. In the case of a 2nd generation poly(amidoamine) (PAMAM) dendrimer terminated with 4-terpyridinyl units, instead of the formation of thermodynamically more stable 2-D arrays, chains that were stacked next to each other ("pearl necklace" formation) were found by STM investigations on highly ordered pyrolytic graphite (HOPG) (Figure 7.1).

Film deposition and complexation were conducted at the phase boundary between a solution of the dendrimer in CH_2Cl_2 and an aqueous solution of $FeSO_4$ on the surface of a freshly cleaved HOPG substrate. The resulting STM images showed a quasi-hexagonal structure in which the intermolecular distances are equivalent in two directions, but different (longer) along the third. As mentioned above, this was explained by the 2-D packing of strands held together by Fe(II) complexation. It is believed that this is the kinetically favored product which would form under these particular applied conditions. Further investigations of these monolayers on HOPG using electrochemical methods, such as cyclic voltammetry (CV) and double potential step chrono-amperometry (DPSCA), were also conducted [6]. By comparing their formal potential ($E^{\circ\prime}$) value of +1.03 V vs sodium saturated calomel electrode (SSCE) to that of free $[Fe(tpy)_2]^{2+}$ ($E^{\circ\prime}$ = +1.10 V vs SSCE), CV experiments indicated that the immobilized species does indeed consist of $[Fe(tpy)_2]^{2+}$ complexes.

Furthermore, by conducting DPSCA measurements, charge propagation dynamics could be studied. For redox-active films, charge propagation can be described as a diffusion process characterized by an apparent diffusion coefficient [7]. By applying the Cottrell equation, values of D_0 could be obtained for dend-8-$[Fe(tpy)_2]^{2+}$ ($5 \pm 2 \times 10^{-7}$ cm^2 s^{-1}) and dend-8-$[Co(tpy)_2]^{2+}$ ($1.3 \pm 0.2 \times 10^{-8}$ cm^2 s^{-1}). These results showed that the electron transfer rate for Fe^{2+}/Fe^{3+} is more than one order of magnitude faster than that of Co^{2+}/Co^{3+}.

A further system consisting of the enantiomeric $[Fe(tpy)_2]^{2+}$ complex deposited as a monomolecular film on HOPG was described by Abruña et al. [8]. As in the case of the dendritic system mentioned above, CV measurements proved the adsorption of a species containing $[Fe(tpy)_2]^{2+}$; the STM images of films derived from each of the metal-enantiomeric ligand complexes showed mirror symmetric structures. The angles of the features observed in relation to the direction of chain propagation were found to be opposite for the two enantiomeric films (Figure 7.2). These observations also match energy-minimized geometries derived from molecular mechanics calculations in terms of structural angles as well as structural length. These observations suggested that each of these entities visualized by STM resembles the upper ligand in a chiral strand, which has no direct contact with the surface. In agreement with the findings from the CD spectra described

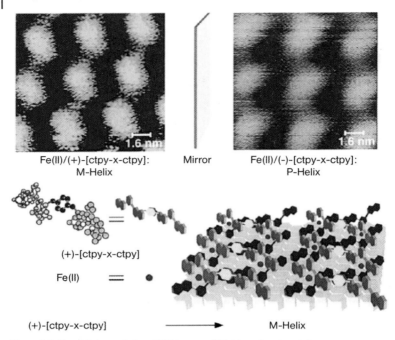

Fe(II)/(+)-[ctpy-x-ctpy]:   Mirror   Fe(II)/(-)-[ctpy-x-ctpy]:
M-Helix   P-Helix

(+)-[ctpy-x-ctpy]

Fe(II)

(+)-[ctpy-x-ctpy]   ⟶   M-Helix

Figure 7.2 Top: High-resolution STM image of highly ordered metallo-supramolecular arrays. Bottom: Energy-optimized structure (MM2) of the (+)-enantiomer and a rendition of the helical structure formed by this bridging ligand and Fe(II).
(Reprinted with permission from [8], © 2001 American Chemical Society).

in Chapter 5, it became evident that films prepared from (+)-[ctpy-x-ctpy] form an M helix and the films prepared from (+)-[ctpy-x-ctpy] form the enantiomeric P helix.

In regards to surfaces other than HOPG, metals such as Pt(100) or Au(111) are widely used as substrates for assembling and investigating monolayer structures. Figgemeier and Constable et al. recently reported the self-assembly of a monolayer of heterogeneous $[Ru(tpy)_2]^{2+}$ or $[Os(tpy)_2]^{2+}$ complexes [9], in which one is unfunctionalized and the other possesses a 4'-(4-pyridinyl)terpyridine (Figure 7.3).

The electron lone pair from the uncomplexed pyridinyl nitrogen should enable adsorption onto a platinum surface. Upon immersing platinum foils or electrodes into an aqueous acetone solution of the complexes for 1 h, spontaneous formation of an adsorbed monolayer was observed. After meticulous rinsing with the solvent, STM as well as electrochemistry techniques were performed. The STM image [Ru(II) complex] showed formation of a hexagonal array with an average distance between adjacent spots of 2.9 nm, while a spot with a radius of ca. 5.5 Å indicated a rather loose packing (surface coverage $= 2.2 \times 10^{-11}$ mol cm^{-2}). CV measurements on both the Ru(II) and Os(II) complexes showed that the peak currents increased linearly with the scan rate, as expected for monolayer formation on the platinum

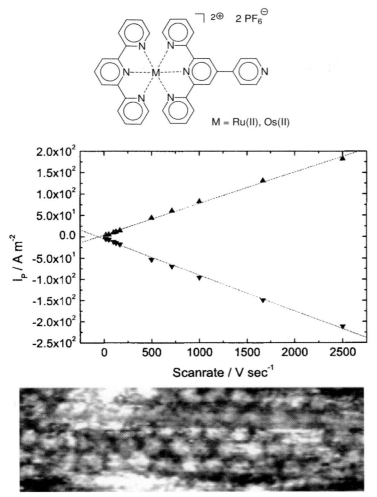

Figure 7.3 Top: Structural image of [M(tpy)(tpy-py)]$^{2+}$, M = Ru(II), Os(II).
Middle: peak current density as a function of scan rate for a monolayer
of [M(tpy)(tpy-py)(PF$_6$)$_2$] on a platinum microelectrode.
Bottom: 14 × 45 nm STM image of a monolayer of [M(tpy)(tpy-py)(PF$_6$)$_2$] on Pt(100).
(Reprinted with permission from [9], © 2003 American Chemical Society).

electrode. Compared with the theoretical value of 90.6 mV for $\Delta E_{p,1/2}$, calculated
for a one-electron process, this value was much lower than the measured values
for either the Ru(II)- or Os(II)-complex monolayers of up to 160 mV. This also
indicated the formation of a loosely packed monolayer. Comparing the saturation
surface coverage derived from the charge under the oxidation and reduction peaks,
a value of 2.5 (± 0.2) × 10^{-11} mol cm^{-2} was obtained, which is in good agreement
with the STM surface coverage of 2.2 × 10^{-11} mol cm^{-2}, as noted above.

Since thiols or disulfides are known to be active precursors for such sulfur-gold bonding, the immobilization of ordered $[M(tpy)_2]^{2+}$ complexes on gold surfaces via the well-known strong sulfur-gold interaction has been demonstrated. Otsuki et al. designed an $[Ru(tpy)_2]^{2+}$ complex, which is functionalized on each of the 5- and 5″-positions with either 4-thiolphenylethenyl or 4-acetylthiophenylethenyl substituents [10]. These ligands are fully conjugated, so that interaction between the surface and the photo- and redox-active Ru(II) was theoretically possible; however, when depositing the complexes on a Au(111) surface, little organization was observed. CV measurements on a gold electrode that was treated with one of these complexes showed that upon sweeping to a negative potential, only a broad desorption peak at a higher voltage around –0.9 to –1.0 V vs Ag/AgCl was observed contrasting with the sharper peak observed from a SAM (self-assembled mono-layer) of octanethiol (–1.02 V).

Another approach with a more complex system was conducted by Kern and Sauvage et al., in which they reported the deposition of a copper catenane on Au(111). Their system comprised one catenane ring containing a terpyridine and a phenanthroline (phen) moiety on the opposite side of the ring, the other ring possessing a phenanthroline and an opposing disulfide moiety to be used for surface binding (Scheme 7.1) [11].

The ambivalence of the coordination state of the copper center [Cu(I) ↔ Cu(II)] has already been proven to be useful in similar systems, e.g., molecular motors and muscles [12, 13]. Here, the investigation of both the Cu(I)- and the Cu(II)-catenane by CV in solution revealed complementary behavior, showing that Cu(I)-catenane adopted a tetracoordinated state whereas the Cu(II)-catenane was pentacoordinated. This indicated that both rings are rotating. Although these observations could not be made for CV measurements on the gold surface, it was shown that the complex was attached to the gold surface by initially opening the disulfide followed by binding to the surface. Again, a difference between oxidation and reduction peaks of less than 60 mV and no broadening of ΔE_p was observed

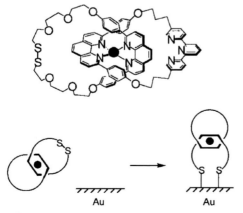

Scheme 7.1 Top: Heteroleptic Co(I) catenane. Bottom: Schematic representation of the adsorption process. (Reprinted with permission from [11]).

when the potential sweep rate was increased. Additionally, polarization modulation-infrared reflection absorption spectroscopy (PM-IRRAS) revealed information about the relative orientations of the interlocked rings, which were both found to be roughly perpendicular to the Au surface.

Apart from directly adsorbing molecules and layers of molecules on surfaces and investigating, a recent example of polyelectrolyte multilayer formation should be mentioned. Kurth et al. investigated alternating layers of poly(styrenesulfonate) and a metallo-supramolecular coordination polyelectrolyte (Co-MEPE) on a poly(ethylenimine) (PEI) modified quartz substrate (Figure 7.4) [14]. The layers $(PSS/Co-MEPE)_n$ (n = 1–10) were characterized by UV/Vis spectroscopy, micro-gravimetry, CV, and permeability and polarity measurements.

UV/Vis spectroscopy indicated that a linear increase for all of the complex bands after each double layer addition was observed. Evidence for multilayer formation was also found from CV data, which showed that the anodic and cathodic current peaks rise proportionally to the square root of scan velocity. This was in contrast to solution measurements, from which a linear rise was observed. Such behavior is characteristic for layers of electrochemical sites, which possess a semi-infinite electrochemical charge diffusion condition. It was further shown by comparative CV measurements that a redox active probe ($[Fe(CN)_6]^{3-}/[Fe(CN)_6]^{4-}$) diffused mostly radially through the layers, which was explained by the more hydrophobic nature of the PSS/Co-MEPE compared to strong polyelectrolyte layers. The same type of system was also investigated using Fe(II) as the metal center, which binds more strongly than Co(II). Here, it was shown that the thicknesses of single films, measured in air by surface plasmon spectroscopy, were 18 ± 3 Å for the coordination polyelectrolyte and 17 ± 2 Å for the PSS, respectively. Thickness measurements at the water-substrate interface revealed a minimum thickness of 25 ± 2 Å for the Fe(II) coordination polyelectrolyte and 27 ± 2 Å for PSS, which can be explained by water penetration of the films. Further examples of polyelectrolyte assembly and terpyridines have been described elsewhere [4, 15–17].

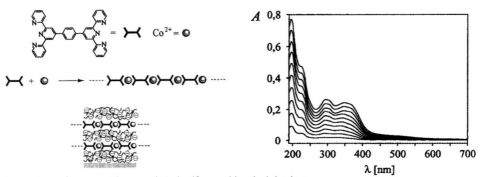

Figure 7.4 Left top: Metal-ion mediated self-assembly, which leads to a coordination polyelectrolyte (Co-MEPE). Left bottom: Multilayer formation by layer-by-layer self-assembly of positively charged Co-MEPE and negatively charged PSS. Right: UV/Vis increase after each double-layer addition. (Reproduced with permission from [14], © 2002 Elsevier).

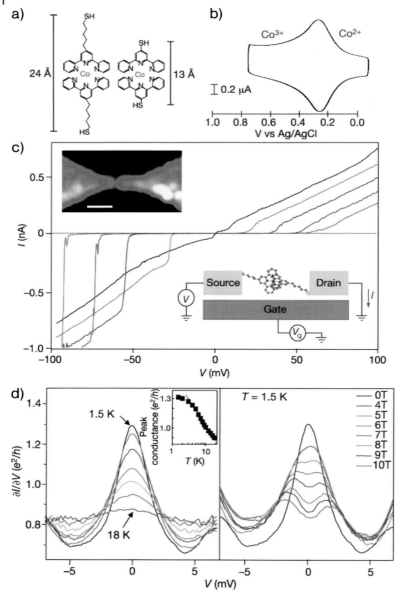

Figure 7.5 (a) Thiol-modified Co(II) complexes and (b) their CV in 0.1 M n-Bu$_4$NPF$_6$/MeCN; (c) I-V curves of a [Co(tpy-(CH$_2$)$_5$-SH)$_2$]$^{2+}$ at different gate voltages (V$_g$) from −0.4 V (red) to −1.0 V (black) with ΔV$_g$ ≈ −0.15 V.
Upper inset: Topographic AFM image of the electrodes with gap (scale bar, 100 nm).
Lower inset: A schematic diagram of the device.
(d) Differential conductance against V showing temperature and magnetic field dependence of the Kondo peak.
(Reprinted by permission from Macmillan Publishers Ltd, Nature [18], © 2002).

Together, these examples demonstrated the increasing control that can be obtained in constructing layered structures containing the $[M(tpy)_2]^{2+}$ motif. Apart from layer formation and stabilization, there is also an increasing interest in controlling and using single molecules or strands of molecules for different applications. In 2002, Park and Pasupathy et al. reported the use of thiol-functionalized $[M(tpy)_2]^{2+}$ complexes, which were introduced into a gap in a 200 nm wide gold wire. The gap produced electromigration on ramping to large voltages at cryogenic temperatures [18]. During this process, some of the complexes, which were initially bound to the gold wire before breakage, migrated to the 1–2 nm gap. Utilizing complexes with different thiol-to-thiol lengths showed different physical effects on a molecular level (Figure 7.5).

For the "longer" complex, behavior corresponding to that of a molecular single-electron transistor was observed, the gate being the degenerately doped Si substrate. For the "shorter" complex, stronger coupling between the ion and the electrons was observed, thus leading to Kondo-assisted tunneling, which can be described as the formation of a bound state between a local spin in the Co(II) center and the conduction electrons in the electrodes, leading to an enhancement of conductance at low biases.

Gold nanoparticles were also modified with thiol-containing terpyridines (Figure 7.6). An example, from 2002, showed the attachment of terpyridinyl thiols (with butyl, octyl, and undecyl spacers) to gold nanoparticles [19]; subsequent addition of Fe(II) ions led to aggregation of the nanoparticles. This concept was further extended to the attachment of Ru(II) complexes bearing a ferrocene to the surface of the gold nanoparticles [20].

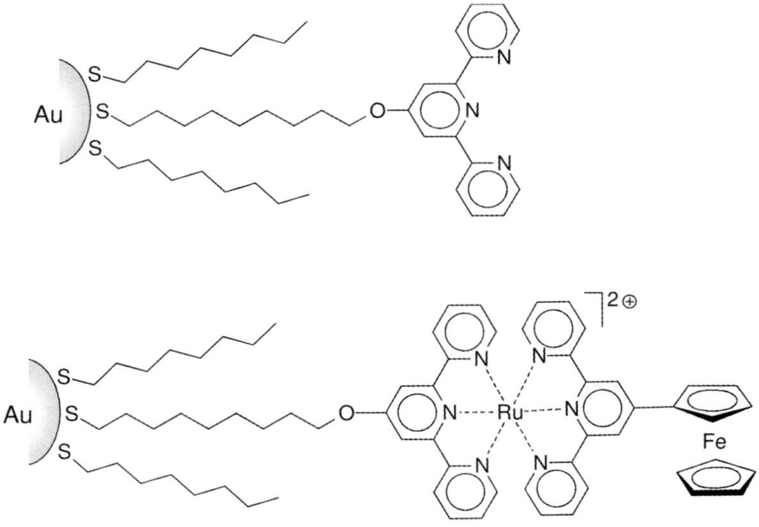

Figure 7.6 Gold nanoparticles, functionalized with terpyridines and ferrocenyl $[Ru(tpy)_2]^{2+}$ complexes [19, 20].

Gaub and Schubert et al. showed that it was possible to determine binding strengths of $[M(tpy)_2]^{2+}$ complexes at the molecular level using single-molecule force spectroscopy [21]. For this purpose, a poly(ethylene glycol) spacer with end-caps of $[Ru(4'-R-tpy)(Cl)_3]^{2+}$ and carboxylic acid was attached to an amino functionalized substrate (Figure 7.7). The same uncomplexed ligand was attached to the tip of the AFM, and upon bringing it into close proximity to the surface, an $[Ru(tpy)_2]^{2+}$ complex was formed. Force extension curves were measured upon

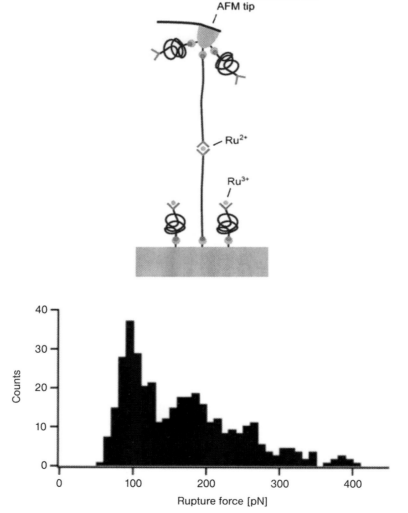

Figure 7.7 Left: Schematic drawing of the experimental set-up.
Right: Histogram of the bond rupture forces (pulling velocity = 118 nm s^{-1}).
(Reprinted with permission from [21]).

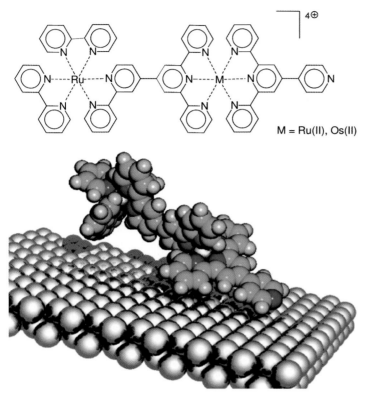

4⊕

M = Ru(II), Os(II)

Figure 7.8 Chemical structure of the dyad and proposed assembly to the platinum surface by molecular modeling [22].
(Reprinted with permission from [22], © 2004 American Chemical Society).

pulling back the tip until the rupture occurred [the experiments were performed under mild conditions in water/DMSO at room temperature to avoid any breaking of the poly(ethylene glycol) chains]. This gave a binding force of around 95 pN for a single complex. The histogram of the bond rupture forces in Figure 7.7 additionally showed weaker peaks at 171 and 253 pN that could be attributed to the rupture of two or three parallel complexes, respectively, by comparison to superimposed theoretical force versus extension curves. It should be possible to expand this technique to similar systems, and this method therefore represents an approach to the characterization of binding strengths of such systems.

Constable et al. have reported *tris*bipyridine/*bis*terpyridine complexes composed of homometallic Ru(II) and heterometallic Ru(II)/Os(II) dyads, which were assembled as monolayers onto a platinum surface (Figure 7.8) [22]. It was shown by fast-scan electrochemistry that these complexes showed a stronger interaction within the monolayer and that the molecules are oriented in a tilted fashion.

7.3
Surface Catalysts

An important application of terpyridine complexes on surfaces that does not necessarily depend on high ordering at the molecular level is that of electrocatalysis. The first work in this direction, to the best of our knowledge, was presented by Meyer et al. They transferred the use of Ru(IV)-containing complexes, as oxidants for various organic substrates in solution, to the electrode surface, using an Ru(IV)-bipyridine-poly-4-vinylpyridine [23]. The same research group bound $[Ru(tpy)(bpy(PO_3H_2)_2)(H_2O)]^{2+}$ to thin films of TiO_2 on glass by phosphonate binding [24]. Upon oxidation with a Ce(IV) derivative, the formation of the catalytically active glass/TiO_2–Ru(IV)=O^{2+} was observed. Test oxidations of cyclohexene, benzyl alcohol, phenol, and *trans*-stilbene revealed a preferred oxidation to the compound that did not exceed the 2-electron level, because the surface binding of the oxidant prevented the formation of lower oxidation states than Ru(II). The dominant product of the oxidation of cyclohexene, for example, was the alcohol and not the ketone; whereas, in the case of *trans*-stilbene, the epoxide was formed, demonstrating the possibility of limiting reductions to a two-electron process through attaching an oxidant to a surface.

The first experiments using ruthenium as a reduction catalyst were performed by Chardon-Noblat et al. using an $[Ru(tpy)(CO)]_n$ complex [25]. The exact structure of this ruthenium-complex polymer is yet unknown, but, based on standard characterization techniques, the existence of metal-metal binding units was confirmed. This material was obtained by electroreduction of isomers of $[Ru(tpy)(CO)Cl_2]$ on the platinum or glassy carbon working electrode surface (Figure 7.9).

Electrocatalysis performed in pure water containing 0.1 M $LiClO_4$ at −1.30 V versus Ag | AgCl on a carbon felt modified with $[Ru(tpy)(CO)]_n$ produced CO and formate with 60 and 15% yield, respectively, after 20 Coulombs had been passed through the solution; however, $[Ru(tpy)(CO)]_n$ was shown to be less stable than $[Ru(bpy)(CO)_2]_n$ under identical experimental conditions. No significant difference between films prepared from the *trans*- and those prepared from the *cis*-complexes was observed.

Regarding the osmium group homolog, an application for the use of PVI-$[Os(dmebpy)(tpy)]^{2+/3+}$ [PVI = poly(N-vinyl imidazole), dmebpy = 4,4′-dimethyl-2,2′-bipyridine], as redox mediator in lacasse-modified electrodes, was recently

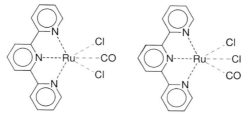

Figure 7.9 Structural formulae of the trans- (left) and cis- (right) [Ru(tpy)(CO)Cl₂] complexes [25].

reported by Barton et al. [26]. Lacasses are polyphenol oxidase enzymes which catalyze the oxidation of a variety of inorganic and aromatic compounds, particularly phenols, with concomitant reduction of molecular oxygen to water. For further detailed results, the reader should consult the literature [26].

Two examples of electrocatalytic reduction with other metals included in $[M(tpy)_2]^{2+}$ complexes, currently exist. Abruña et al. found that electropolymerized films of vinyl-terpyridine complexes of iron, nickel, and cobalt were effective catalysts for CO_2 reduction as well as the two- and four-electron reduction of oxygen [27]. These films were found to be quite robust to washing with different solvents and to continuous potential cycling, the latter leading to less than 10% loss after 6 h. The catalytic activity observed for the films was higher than that for the free complexes in solution. Films of $[Ni(vinyl-tpy)_2(PF_6)_2]$ were found to catalyze both two- and four-electron reduction of oxygen, which does not occur when using $[Ni(tpy)_2(PF_6)_2]$ in homogeneous solution. The reason for these findings is believed to lie in cooperativity effects between the metal centers, which are more closely packed in films than they are in solution. Regarding the differences originating from the use of different metals, Co exhibited the highest activity, probably because it acts as a two-electron donor through Co(III/I) couples; this is not possible for either iron or nickel.

In a similar approach, Elliott et al. used *bis*-terpyridine ligands, which, upon complexation, resulted in oligomeric and/or polymeric materials [28]. The polymerization was conducted directly on a glassy carbon electrode, resulting in film-coated electrodes. In contrast to the results for $[metal(vinyl-tpy)_2(PF_6)_2]$ films described above, there was no improvement in electroreduction compared to the free metallo-polymer in solution. The authors attributed this fact to the rigidity of the latter system, which possibly does not allow a coordination site to open in order to become catalytically active, which was observed for $[Co(vinyl-tpy)_2(PF_6)_2]$ films in MeCN. Upon reduction from Co(II) to Co(I), the terpyridine ligands are partially and reversibly displaced by two MeCN molecules. The examples mentioned here demonstrate the feasibility of terpyridine complexes to act as redox-active materials.

7.4
Photoactive Materials

Another major area of research incorporating terpyridine complex surface interactions is that of interfacial photophysical processes, especially systems concerning the conversion of solar light to energy. First examples include photoelectrodes based on an electropolymerized molecular ruthenium dyad reported by Collin et al. [29] in which polymer films incorporating a molecular dyad of the type $<V^{2+}-[Ru(II)-(ptpy)_2]^{2+}>$ (V = methylviologen, ptpy = 4′-phenyl-terpyridine) (Figure 7.10, top) were described. Thin films were prepared by anodic electropolymerization of the pyrrole groups on the ligand opposite to the ligand containing the methylviologen on an indium tin oxide (ITO) electrode.

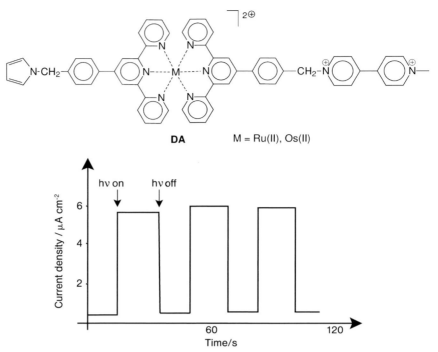

Figure 7.10 Top: Donor-acceptor system **DA** with a pyrrole (anchor), an $[Ru(tpy)_2]^{2+}$ or $[Os(tpy)_2]^{2+}$ complex (photochemical center), and a methylviologen (MV^{2+}, donor). Bottom: Photocurrent response of an ITO/poly-(**DA**) modified electrode. (Reprinted with permission from [29], © 1991 American Chemical Society).

Upon irradiation using visible light and in the presence of *tris*(ethanol)amine (TEOA, irreversible electron donor), an anodic photocurrent was observed when the electrode was potentiostatted to 0 V. The photoactive center was first excited by visible light, upon which the charge-separated state with the methylviologen was formed. Ruthenium(III) then irreversibly oxidized TEOA, and the photocurrent was produced by electron transfer into the polymer and to the electrode. The steady-state photoresponse was moderately stable with time (loss of 19% after 30 min) in accordance with the stability of the modified electrode (Figure 7.10, bottom). The maximum steady-state current I_s for different surface concentrations Γ was found to be 8 µA cm^{-2} at $\Gamma = 6 \times 10^{-10}$ mol cm^{-2}. These results are in good agreement with results obtained for a comparable $<V^{2+}$-$[Ru(bpy)_3]^{2+}>$ system, except that the limiting value of Γ was lower for the terpyridine case.

These results can be compared to the work conducted by Collin et al., where copolymers containing monomers of $[Ru(ttpy)]^{2+}$ (photoactive center) and monomers of a viologen (electron acceptor) were used [30]. For this system, in which the photoactive center and viologen are not directly covalently linked but only through the polymer backbone, only a moderate maximum I_s of < 1.5 µA

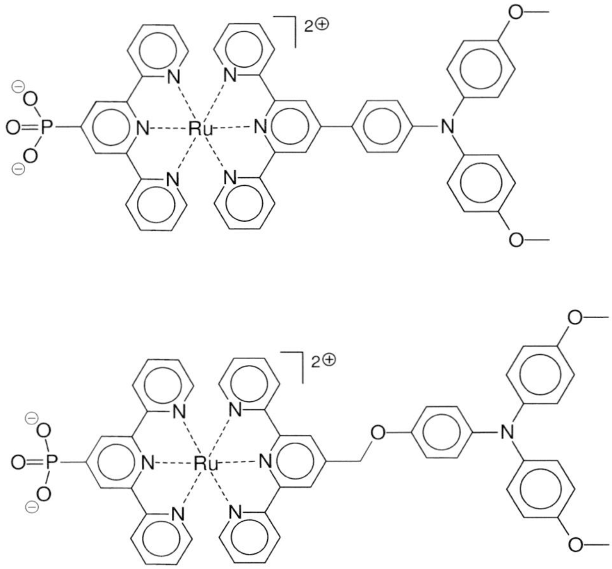

Figure 7.11 Bonhôte's electron donor-sensitizer dyads [32, 33].

cm^{-2} was found, indicating that intramolecular photoelectron transfer was more efficient than intermolecular transfer.

Another example of an electropolymerized system was shown by Hanabusa et al. Using the same basic approach, an aniline-modified $[Ru(tpy)_2]^{2+}$ complex was electropolymerized via a reaction mechanism similar to the electropolymerization of aniline [31]. The photocurrent response of this polymer film on an ITO electrode in aqueous 0.1 M $LiClO_4$ solution containing 0.5 mmol of methylviologen (MV^{2+}) and saturated with oxygen showed a maximum photocurrent of 1.9 $\mu A\ cm^{-2}$. The sample was irradiated with a 300-W halogen lamp. Because of rapid oxidation of $MV^{+\cdot}$ with oxygen, the unfavorable reverse electron reaction was suppressed.

Further progress in this area was made by choosing nanocrystalline, semi-conducting metal oxides, such as TiO_2 (anatase modification) on conducting glass, as the electrode. Bonhôte et al. investigated charge separation on a donor-sensitizer system (Figure 7.11, top) [32, 33]. The observed charge separation was, however, not found to be more efficient than that of a model sensitizer without the donor system because of the very fast recombination reaction. For a similar system (Figure 7.11, bottom) with a different spacer between donor and sensitizer, the photoinduced charge separation was found to be more efficient, because the electron in the excited state was localized more on the ligand bound to the semiconductor surface.

More recently, Grätzel et al. synthesized the black dye $[RuL(NCS)_3(HNEt_3)]$ (L = tpy), which exhibits a larger red shift in absorption than that of other compounds from the $[Ru(tpy)_2]^{2+}$ family (Figure 7.12) [34, 35]. A photovoltaic cell consisting of an adsorbed monolayer on TiO_2 conducting glass in conjunction

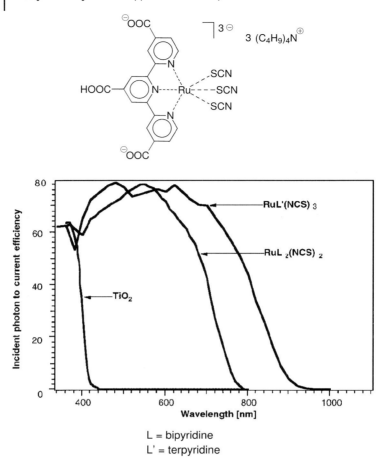

L = bipyridine
L' = terpyridine

Figure 7.12 Top: Salt of the $[Ru(4,4',4''-(^-O_2C)_3-tpy)(SCN)_3]$ panchromatic sensitizer.
Bottom: Its photocurrent action spectrum comparison with the spectra of the
$[Ru(4,4'-(^-O_2C)_2-bpy)_2(SCN)_2]$ analog and bare TiO_2 [34].
(Reprinted with permission from [35], © 2001 American Chemical Society).

with the redox electrolyte LiI/LiI_3 in propylene carbonate showed an incident
photon-to-current conversion efficiency (IPCE) of 80% over a broad region with a
photocurrent density of 20.5 mA cm^{-2}. Under standard AM 1.5 sunlight (air mass
1.5 sunlight is the spectrum of sunlight that has been filtered by passing through
1.5 thicknesses of the earth's atmosphere), an open circuit voltage of 0.72 V was
observed, which gave an overall conversion efficiency of over 10%.

This improved system was compared with a bipyridine analog, which had been
up to that point the best-performing charge-transfer sensitizer (~18 mA cm^{-2}).
Following that discovery, other research projects were conducted including
adsorption and crystallization studies [36] and investigations of interfacial electron-
transfer dynamics [37]. In an attempt to improve the molar extinction coefficient,

Huang et al. investigated the analogous 4′-carboxyphenyl-substituted compound, which, probably because of the absence of two carboxy groups on the outer rings, showed a worse performance in a photovoltaic cell [38]. Sugihara and Arakawa et al. tested another analogous compound, which incorporated a long alkyl chain on the β-diketonato ligand [39]. Under similar conditions, the photocurrent action spectrum showed behavior similar to that of the mother complex described above. In accordance with the absorption spectra, the alkyl complex showed higher incident photon-to-current efficiency (IPCE) values (in the 720–900 nm region) than the parent. By comparing the photocurrents in the presence and absence of deoxycholic acid, an indication that the long hydrocarbon chain prevented surface aggregation of the sensitizer was found.

Apart from solar cell research, two other photophysically interesting examples have been reported: (a) photodegradation of organic material and (b) organic light-emitting devices (OLEDs). Lam et al. reported on the TiO_2 photodegration of CCl_4 in aqueous medium (Figure 7.13) [40]. The $[Ru(4'-(4-H_2O_3PC_6H_4)-tpy)_2(PF_6)_2]$ complex (Figure 7.13, top) was mixed with a suspension of indium tin oxide (ITO), and, after filtration, drying, and grinding, the resulting powder was used for degradation experiments. A 100-W tungsten lamp was used for photolysis in a CCl_4-saturated aqueous solution containing potassium iodide as a reductant. Figure 7.13 (bottom) shows the principle of the catalytic cycle. The rate of degradation was found to obey a Langmuir-Hinshelwood rate law, which is expected because only those CCl_4 molecules that are adsorbed onto the surface of TiO_2 can be reduced.

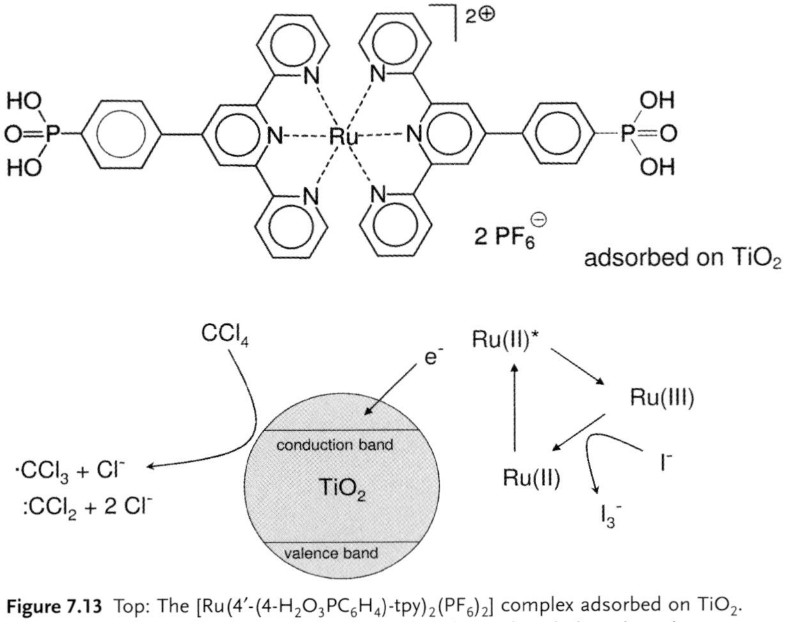

Figure 7.13 Top: The $[Ru(4'-(4-H_2O_3PC_6H_4)-tpy)_2(PF_6)_2]$ complex adsorbed on TiO_2. Bottom: Schematic representation of the visible light mediated photodegradation of CCl_4 in aqueous medium [40].

Elliott et al. investigated an electropolymerized $[Ru(tpy)_2]^0$ film, which was vapor deposited onto an $Alq_3/TPD/ITO$ [$Alq_3 = tris$(8-hydroxyquinoline) Al(III) complex, emissive and electron transport layer/TPD = triarylamine derivative, hole transport material] substrate in order to create an electroluminescent device [41]. The $[Ru(tpy)_2]^0$ film acts as a low-work-function electron-injecting contact and is of special interest because it can be deposited by thermal evaporation, making it easy to handle and to control. OLEDs were constructed using $[Ru(tpy)_2]^0$ as low-work-function organic material (LWOM) and the architecture Ag/LWOM/ Alq_3(400 Å)/TPD(400 Å)/ITO. Figure 7.14 shows the performance of such a device. Upon reducing the barrier-to-hole injection between ITO and TPD by including the conducting polymer interlayer poly(3,4-ethylenedioxythiophene)-polystyrene-sulfonic acid (PEDOT-PSS), a significant improvement was observed. When more current was used, more light was produced at a given voltage. The metal layer covering the LWOM was also varied from Ag to Au, which has a very high work function (5.2 eV). Devices with different metal layers showed comparable performance, thus indicating that the nature of the metal contact was of secondary importance to the underlying LWOM. This should allow for variation of the metal contact, leading to devices which are, for example, more oxidation resistant and which therefore lead to increased device lifetimes.

An example of a blue LED was shown recently by Che et al. [42] in which they used a $[Zn(tpy)_2]^{2+}$ polymer spin-coated on ITO with the device structure, e.g., $ITO/PEDOT:PSS/[Zn(tpy)_2]^{2+}$-polymer/Ca/Al. A peak maximum of 450 nm was observed in the electroluminescence spectrum, with the blue EL intensity

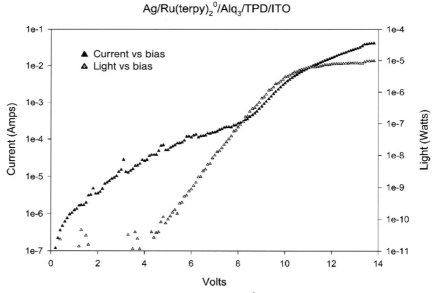

Figure 7.14 Performance of an OLED with an $[Ru(tpy)_2]^0$/Ag cathode. (Reprinted with permission from [41], © 2003 American Chemical Society).

increasing with increasing bias voltage. This example again demonstrated the versatility of these $[M(tpy)_2]^{2+}$ complexes, leading to, in this case, encouraging results for the search for a stable and intense blue light emission, which is currently of particular interest for many photo-optical applications.

References

1 J. P. Spatz, *Angew. Chem. Int. Ed.* **2002**, *41*, 3359–3362.
2 S. Krämer, R. R. Fuierer, C. B. Gorman, *Chem. Rev.* **2003**, *103*, 4367–4418.
3 S. Welter, K. Brunner, J. W. Hofstraat, L. De Cola, *Nature* **2003**, *421*, 54–57.
4 T. Salditt, U. S. Schubert, *Rev. Mol. Biotechnol.* **2002**, *90*, 55–70.
5 D. J. Diaz, G. D. Storrier, S. Bernhard, K. Takada, H. D. Abruña, *Langmuir* **1999**, *15*, 7351–7354.
6 D. J. Diaz, S. Bernhard, G. D. Storrier, H. D. Abruña, *J. Phys. Chem. B* **2001**, *105*, 8746–8754.
7 D. A. Buttry, F. C. Anson, *J. Am. Chem. Soc.* **1983**, *105*, 685–689.
8 S. Bernhard, K. Takada, D. J. Diaz, H. D. Abruña, H. Murner, *J. Am. Chem. Soc.* **2001**, *123*, 10265–10271.
9 E. Figgemeier, L. Merz, B. A. Hermann, Y. C. Zimmermann, C. E. Housecroft, H. J. Guentherodt, E. C. Constable, *J. Phys. Chem. B* **2003**, *107*, 1157–1162.
10 J. Otsuki, H. Kameda, S. Tomihira, H. Sakaguchi, T. Takido, *Chem. Lett.* **2002**, 610–611.
11 L. Raehm, J.-M. Kern, J.-P. Sauvage, C. Hamann, S. Palacin, J.-P. Bourgoin, *Chem. Eur. J.* **2002**, *8*, 2153–2162.
12 B. X. Colasson, C. Dietrich-Buchecker, M. C. Jimenez-Molero, J.-P. Sauvage, *J. Phys. Org. Chem.* **2002**, *15*, 476–483.
13 J.-P. Sauvage, *Chem. Commun.* **2005**, 1507–1510.
14 D. G. Kurth, M. Schütte, J. Wen, *Colloids Surf., A* **2002**, *198–200*, 633–643.
15 Y. Bodenthin, U. Pietsch, H. Möhwald, D. G. Kurth, *J. Am. Chem. Soc.* **2005**, *127*, 2110–3114.
16 D. G. Kurth, J. P. López, W.-F. Dong, *Chem. Commun.* **2005**, 2119–2121.
17 J. P. López, W. Kraus, G. Reck, A. Thünemann, D. G. Kurth, *Inorg. Chem. Commun.* **2005**, *8*, 281–284.
18 J. Park, A. N. Pasupathy, J. I. Goldsmith, C. Chang, Y. Yaish, J. R. Petta, M. Rinkoski, J. P. Sethna, H. D. Abruña, P. L. McEuen, D. C. Ralph, *Nature* **2002**, *417*, 722–725.
19 T. B. Norsten, B. L. Frankamp, V. M. Rotello, *Nano Lett.* **2002**, *2*, 1345–1348.
20 T.-Y. Dong, H.-W. Shih, L.-S. Chang, *Langmuir* **2004**, *20*, 9340–9347.
21 M. Kudera, C. Eschbaumer, H. E. Gaub, U. S. Schubert, *Adv. Funct. Mater.* **2003**, *13*, 615–620.
22 E. Figgemeier, E. C. Constable, C. E. Housecroft, Y. C. Zimmermann, *Langmuir* **2004**, *20*, 9242–9248.
23 G. J. Samuels, T. J. Meyer, *J. Am. Chem. Soc.* **1981**, *103*, 307–312.

24 L. A. Gallagher, T. J. Meyer, *J. Am. Chem. Soc.* **2001**, *123*, 5308–5312.

25 S. Chardon-Noblat, P. Da Costa, A. Deronzier, S. Maniguet, R. Ziessel, *J. Electroanal. Chem.* **2002**, *529*, 135–144.

26 S. C. Barton, H.-H. Kim, G. Binyamin, Y. Zhang, A. Heller, *J. Phys. Chem. B* **2001**, *105*, 11917–11921.

27 C. Arana, M. Keshavarz, K. T. Potts, H. D. Abruña, *Inorg. Chim. Acta* **1994**, *225*, 285–295.

28 D. L. Feldheim, C. J. Baldy, P. Sebring, S. M. Hendrickson, C. M. Elliott, *J. Electrochem. Soc.* **1995**, *142*, 3366–3372.

29 J. P. Collin, A. Deronzier, M. Essakalli, *J. Phys. Chem.* **1991**, *95*, 5906–5909.

30 J. P. Collin, A. Jouaiti, J. P. Sauvage, *J. Electroanal. Chem. Interfac. Electrochem.* **1990**, *286*, 75–87.

31 K. Hanabusa, A. Nakamura, T. Koyama, H. Shirai, *Polym. Int.* **1994**, *35*, 231–238.

32 P. Bonhote, J.-E. Moser, R. Humphry-Baker, N. Vlachopoulos, S. M. Zakeeruddin, L. Walder, M. Graetzel, *J. Am. Chem. Soc.* **1999**, *121*, 1324–1336.

33 P. Bonhote, J. E. Moser, N. Vlachopoulos, L. Walder, S. M. Zakeeruddin, R. Humphry-Baker, P. Pechy, M. Grätzel, *Chem. Commun.* **1996**, 1163–1164.

34 M. K. Nazeeruddin, Pechy P., Grätzel, M., *Chem. Commun.* **1997**, 1705–1706.

35 M. K. Nazeeruddin, P. Pechy, T. Renouard, S. M. Zakeeruddin, R. Humphry-Baker, P. Comte, P. Liska, L. Cevey, E. Costa, V. Shklover, L. Spiccia, G. B. Deacon, C. A. Bignozzi, M. Grätzel, *J. Am. Chem. Soc.* **2001**, *123*, 1613–1624.

36 V. Shklover, M. K. Nazeeruddin, M. Grätzel, Y. E. Ovchinnikov, *Appl. Organomet. Chem.* **2002**, *16*, 635–642.

37 C. Bauer, G. Boschloo, E. Mukhtar, A. Hagfeldt, *J. Phys. Chem. B* **2002**, *106*, 12693–12704.

38 Z.-S. Wang, C.-H. Huang, Y.-Y. Huang, B.-W. Zhang, P.-H. Xie, Y.-J. Hou, K. Ibrahim, H.-J. Qian, F.-Q. Liu, *Sol. Energy Mater.* **2002**, *71*, 261–271.

39 A. Islam, H. Sugihara, M. Yanagida, K. Hara, G. Fujihashi, Y. Tachibana, R. Katoh, S. Murata, H. Arakawa, *New J. Chem.* **2002**, *26*, 966–968.

40 S. T. C. Cheung, A. K. M. Fung, M. H. W. Lam, *Chemosphere* **1998**, *36*, 2461–2473.

41 C. J. Bloom, C. M. Elliott, P. G. Schroeder, C. B. France, B. A. Parkinson, *J. Phys. Chem. B* **2003**, *107*, 2933–2938.

42 S.-C. Yu, C.-C. Kwok, W.-K. Chan, C.-M. Che, *Adv. Mater.* **2003**, *15*, 1643–1647.

Subject Index

Modern Terpyridine Chemistry. U. S. Schubert, H. Hofmeier, G. R. Newkome
Copyright © 2006 WILEY-VCH Verlag GmbH & Co. KGaA, Weinheim
ISBN: 3-527-31475-X